James Walvin is the author of many books on slavery and modern social history. His book, *Crossings*, was published by Reaktion Books in 2013. His first book, with Michael Craton, was a detailed study of a sugar plantation: *A Jamaican Plantation, Worthy Park, 1670–1970* (Toronto, 1970). He became a Fellow of the Royal Society of Literature in 2006, and in 2008 was awarded an OBE for services to scholarship.

Also by James Walvin:

Slavery in Small Things: Slavery and Modern Cultural Habits

Different Times: Growing Up in Post-War England

Crossings: Africa, the Americas and the Atlantic Slave Trade

The Zong: A Massacre, the Law and the End of Slavery

The Trader, The Owner, The Slave:
Parallel Lives in the Age of Slavery

Atlas of Slavery

Black Ivory: Slavery in the British Empire

Questioning Slavery

An African's Life:
The Life and Times of Olaudah Equiano, 1745–1797

Making the Black Atlantic: Britain and the African Diaspora

HOW
SUGAR
CORRUPTED
THE WORLD

FROM
SLAVERY
TO
OBESITY

JAMES WALVIN

ROBINSON

ROBINSON

First published as *Sugar* in Great Britain in 2017 by Robinson

This paperback edition published in 2019 by Robinson

A CIP catalogue record for this book is available from the British Library

ISBN: 978-1-47213-812-5

Typeset in Adobe Garamond by Hewer Text UK Ltd, Edinburgh
Printed and bound in Great Britain by Clays Ltd, Elcograf S.p.A.

Papers used by Robinson are from well-managed
forests and other responsible sources

Robinson
An imprint of
Little, Brown Book Group
Carmelite House
50 Victoria Embankment
London EC4Y 0DZ

An Hachette UK Company
www.hachette.co.uk

www.littlebrown.co.uk

Contents

Our new grandson, Max Walvin,
arrived when I found myself immersed in sugar;
he quickly proved himself to be the sweetest thing of all.

This book is for him.

Acknowledgements

T HE PERSONAL AND professional debts I owe for my understanding of sugar go back a very long way. I was first
introduced to the history and economics of sugar in the summer
of 1967 when I began work on the papers of Worthy Park
Estate in Jamaica. Over the subsequent fifty years, my friends
in Jamaica have always welcomed me back, providing hospitality, friendship and practical help whenever I asked. I owe special
thanks to Robert and Billie Clarke, and to Peter and Joanie
McConnell at Worthy Park. David and Andrea Hopwood,
Oliver Clarke and Monica Ladd have been equally supportive
and hospitable over the years. I hope they all realise how much
their friendship and support means to me.

In those same years, I have worked on different aspects of the
history of sugar in a number of libraries and archives, in
Jamaica, Barbados and the United Kingdom. Though my interests often strayed into related fields – notably slavery and the
slave trade – the story of sugar has remained central to much of

what I have written and taught. But it was only when I came to write this book that I began fully to appreciate how much my work has been influenced by sugar.

My greatest personal debt is to my late friend Michael Craton who first persuaded me, after meeting in graduate school, to join him in what I initially thought was a speculative venture in Jamaica. The outcome was our book, *A Jamaican Plantation: The History of Worthy Park, 1670–1970* (1970). Working together in Jamaica between 1967 and 1970 opened my eyes not merely to the importance of sugar and slavery, but raised serious questions about how we should view British history more broadly. Throughout, Michael was a demanding mentor, insistent on careful scholarship and, equally, on well-crafted writing. His editorial lessons helped to shape my subsequent writing career.

Gad Heuman, with whom I worked for many years as co-editor of the scholarly journal *Slavery and Abolition,* has been a great supporter and influence throughout – but, better still, a friend. And no one working on the Caribbean can avoid a massive debt to Barry Higman (one of the major historians of his generation) whose remarkable work, of unrivalled range and detail, influences this book throughout.

A starting point for anyone interested in the history of sugar is Elizabeth Abbott's important book *Sugar: A Bittersweet History* (2008). Like all students of sugar, however, my greatest debt is to Sidney Mintz, above all for his remarkable book *Sweetness and Power* (1985). Mintz was a hugely influential scholar, and a wise and gentle critic. He was also a great encourager of younger people. The book that follows is not intended as a successor to his book, but it certainly could not have been written without it.

ACKNOWLEDGEMENTS

Three libraries were especially important in the emergence of this book. The University of York Library, the Swem Library of the College of William and Mary and, above all, the Wellcome Library in London. I am also indebted to James W. Johnson, then President of the United States Beet Sugar Association, who granted me access to the Association's library in their Washington offices.

My work on this book has been made more pleasurable by the hospitality of a number of friends: Martin and Rachel Pick in London; Patsy Sims and Bob Cashdollar in Washington; and Bill and Elizabeth Bernhardt in New York. In Williamsburg, over many years, I have been fortunate in my friendship with Marlene and Bill Davis, and Tolly and Ann Taylor – all of whom opened their homes to me. I am immensely grateful to Ben Hayes who provided the initial encouragement to write this book. Charles Walker, my agent, was again supportive throughout, as was my editor, Duncan Proudfoot; and my thanks, too, to Jon Davies, whose copy-editing has greatly improved the final version of this book.

Jenny Walvin, as always, makes everything possible.

Preface

THE SHOP DIRECTLY opposite my childhood home was a small newspaper shop. But all the neighbourhood children called it 'the sweet shop'. The counter was covered with a variety of national and local newspapers, but behind the counter were stacked rows of bottles and jars filled with the sweets we loved and which we were given as treats or bought with our spare pennies – as much, that is, as the ration books in the 1940s and 1950s allowed. For most of the time, however, we could only look at the jars enviously; money and rations were in short supply. Fifty yards away up the street, there was another sweet shop, in a small dowdy bungalow, which sold nothing but sweets and chocolates. As if that were not enough, we could stride back across the street to the local Co-op with its own enticing offerings of chocolates and sweets (and biscuits and cakes as well). Even in those straitened times, all this seemed a cornucopia of sweet things, and all within a mere hundred yards of our front door.

We all had a sweet tooth, and the constraints of wartime and post-war rationing only made the craving worse. Sometimes we bartered one rationed item for another, swapping life's essentials for a sweet pleasure. Our mother once exchanged our bacon ration for our grandparents' sweets ration.

This addiction to sweets and chocolates wasn't just a family matter, but was deeply entrenched in the whole community; all my childhood friends and their families were equally addicted. On high days and holidays, birthdays, Christmas and Whitsuntide (a major event in Manchester), children were treated to special gifts of chocolates and sweets. Even on our summer seaside holidays – that annual Lancastrian trek from the cotton towns to the Irish Sea – one seaside treat was tackling lengths of that sticky, tooth-defying, sugar-filled Blackpool rock. Predictably, when sweet rationing finally ended in 1953, the local shops were swiftly cleaned out of the sweet temptations we had all looked forward to – my brother and I managed to get our hands on a small box of Cadbury's Milk Tray.

Our love of sugary treats was only one example of the role sugar played in our lives. In fact, sugar was everywhere. It held pride of place, alongside the teapot, on the kitchen table that served as a gathering point not merely for meals but as a rendezvous for the gaggles of women who trooped in and out of the house throughout the day. Daily social life was lubricated by regular servings of sweet tea. My grandpa could have rivalled Dr Johnson in his love of strong tea; he always had a half-pint mug of strong tea to hand, and it, too, was sweetened with lashings of sugar, scooped from the bag that lived permanently on the table in the room that served as kitchen, dining and sitting room.

My mother and her women friends, like the nation at large, moaned endlessly about shortages, but especially about the scarcity of sugar, although what they were allowed now seems lavish in retrospect – far, far more than my own family would use in the course of a week. But at that time, from 1942–53, sugar was added to everything. It even accompanied us to school. We were dispatched to primary school with a snack for the morning break: slices of toast or bread, glued together with heavily sweetened jam, or simply sprinkled with sugar.

All this took place in a society that was severely rationed and we all ticked along in pinched conditions, making do as best we could. Yet throughout, sugar was everywhere. It was (like tobacco smoke) an inescapable fact of life, and so integral to the way we all lived that we did not even notice it – except, that is, when we ran short.

We were also regular visitors to the local dentist. Not for regular check-ups, but to remove the damage wrought by our sugary diet. All my older relatives had dentures. My father had all his teeth removed at the age of twenty-one; my mother lost her remaining teeth in her mid-thirties. Grandma, uncles, aunts and close family friends – all had dentures. Grandpa was an exception; his few surviving teeth were like Elizabeth I's – gnarled and discoloured – but they just about served their purpose. No one thought it odd or unusual to be without their own teeth, even at an age which, today, seems shockingly young. Teeth were extracted in people's early years, partly for financial reasons – it was cheaper to have them whipped out than to spend money on regular dental visits. Overwhelmingly, however, teeth were extracted because they were rotten.

In my family – and I suspect throughout my whole community – false teeth were more common than healthy, natural teeth among adults. Dentures were even a cause of family hilarity. One set shot out when a relative sneezed. When relatives were confined to bed, I recall visiting them and being transfixed by their dentures grinning at me from a bedside glass of water. When one elderly neighbour lost his teeth, we all turned his home upside down to look for them – in vain. On more formal family occasions – those Sundays when invited to 'tea' – relatives' ill-fitting, clicking dentures were a giveaway, clues to a much wider, more significant narrative. These personal recollections of family life are important elements in the story that follows. Of course, I didn't realise it at the time, but it now seems obvious. What lies behind all this is the story of the widespread damage and corruption caused by sugar.

It took a very long time for the penny to drop. Even when I lived and worked on a Jamaican sugar estate in the late 1960s, I didn't think about the connection between sugar and the health of people who consumed it. As a newly minted academic historian, I was hard at work on what became, in league with a friend and colleague, my first published book: the history of one sugar plantation between 1670 and 1970. It was studying Jamaica's sugar fields that set me on course for my subsequent academic career as a student of slavery. But in the beginning, I didn't make the connection between Africans in Jamaican sugar fields and the world I grew up in, in the north of England. Yet both were intimately linked.

We have come to think of sugar very differently in the early twenty-first century, and the book that follows is an attempt to explain how that happened. At one level, we view it differently partly because we know so much more about it. But sugar has also taken a route no one could have predicted, even a generation ago. Very few people suggested, say in 1970, that sugar posed a global health problem. Yet, today, sugar is regularly denounced as a dangerous addiction – on a par with tobacco – and is the cause of a global epidemic of obesity.

But how did this come about? How did a simple commodity that was once a prized monopoly of kings and princes become an essential ingredient in the lives of common people – before mutating yet again into the apparent cause and occasion of major global health problems?

Introduction

Sugar in Our Time

H OW DID IT come to this? What persuaded tens of millions of people the world over to like – to *need* – a commodity, sugar, which medical science now insists is bad for us? As if to compound the confusion, in the summer of 2016 we were bombarded by adverts proclaiming a product because it contained *no* sugar. That summer, millions of TV viewers were exposed to a very unusual advertisement for Coca-Cola. At matches played at football's Euro finals in France, and watched by millions globally (the entire competition partly sponsored by Coca-Cola), adverts flashed along the electronic hoardings telling us that their new drink contained 'Zero Sugar'. Anyone watching a game would have seen that message – 'Zero Sugar' – dozens of times.

Those games were, of course, an excellent platform for adverts. Next to the Olympics and the World Cup, the Euro finals were guaranteed to generate a global audience counted in the hundreds of millions. But what was striking about this particular advert was that it was promoting a product by asserting what it did *not* have; it was announcing a drink that *lacked* something, a drink that did *not* contain sugar. Launching that product had been a costly business – £10 million in the UK alone.[1] It is hard to think of a comparable promotion – lauding a product, not for what it offers, but for what it *doesn't* offer. Here is a drink without sugar.

For English viewers, it might have seemed a timely advert because, only a year earlier, a major Government report had highlighted the problem of sugar-related obesity among millions of English people.[2] Although sugar has been part of our diet for centuries, in recent years it has become a subject of contentious social and political debate. In my own childhood (in the years of wartime and post-war shortages and rationing), my parents often complained about not being able to get enough sugar. Today, parents are discouraged – by doctors, newspapers and politicians – from consuming too much sugar. For centuries, children were pampered and soothed by being given sweet treats; today, the principal drive is to *restrict* children's access to sugar and all sweet things. In fact, sugar has taken on a pariah status. Yet, within living memory, it was widely viewed both as a necessity *and* a pleasurable essential – a commodity that fortified and pleasured in equal part. What has brought about this extraordinary change in the way we see and talk about a commodity that has been part of the human diet for centuries?

* * *

Though part of Western diet for many centuries, before roughly 1600 sugar was a costly luxury, available only to the rich and powerful. All that changed in the course of the seventeenth century, with the rise of European sugar colonies in the Americas. Thereafter, sugar became cheap, ubiquitous and hugely popular. What had formerly been a costly item now became an everyday necessity. Sugar that had once graced only the tables of society's elites was, by 1800, one of life's essentials even for the poorest of working people. And that was how sugar remained, until the mid-twentieth century – an unquestioned part of the lives of millions and a vital ingredient in a wide variety of food and drink. Yet today, when sugar is discussed in the media, it is portrayed as a threat to health – a major contributor not only to individual ill health but also the cause of a global epidemic of obesity. As a result, sugar has become a matter of pressing concern for governments and international health organisations.

Today, people the world over consume sugar in staggering volumes, with consumption highest in countries which produce sugar – Brazil, Fiji and Australia, for example. Australians consume more than 50kg per person each year. But these levels are only slightly lower in other countries – such as in Europe and North America – places that first pioneered mass consumption of sugar after 1600. Yet even these broad generalisations have changed quite dramatically over the past generation, thanks largely to the impact of modern fast foods and fizzy drinks, most of which come laden with sugar. Much of their sweetness today derives, however, not from cane sugar but from corn or chemical sweeteners.

The taste for sweetness in food and drink is universal, and the cultivation of sugar is global. A great variety of sugar cane is cultivated in the tropics, while sugar beet is cultivated in temperate regions. But the engine behind the rise of sugar's popularity was cane sugar. Its early history, in Indonesia, India and China, was small-scale and aimed solely at local markets. But when sugar cane was transplanted to plantations in the Mediterranean, then into islands in the Atlantic, the story changed – and even more dramatically when sugar crossed the Atlantic to the Americas. There, sugar cane was cultivated and converted to sugar by enslaved Africans (themselves shipped across the Atlantic). It was this slave-grown sugar that brought about revolutionary changes in the landscape of the sugar colonies while transforming the tastes of the Western world.

As Europeans and Americans settled and traded with the wider world in the course of the nineteenth century, they transplanted commercial sugar cultivation to new locations: to islands in the Indian Ocean, to Africa, Indonesia, to Pacific islands and to Australia. But wherever sugar took hold, local sugar planters had problems with labour. They found the answer in imported, indentured labour. From one sugar region to another – from Brazil to Hawaii – the sugar plantation became the home of alien people – people who had been uprooted and shipped vast distances to undertake the gruelling, intensive labour on sugar plantations.

For all that, sugar plantations more than proved their worth to their owners and investors. But there was a price to pay for the development of the sugar plantation. The natural environments were hugely damaged by the development of sugar. From Barbados in the 1640s to the Florida Everglades in recent years, the ecological harm caused by sugar plantations has been enormous, and is only now being fully recognized. It is, however, the *human* cost of sugar cultivation which is most obvious and dramatic. It is at its most visible in the labour force, from the first slave gangs in sixteenth-century Brazil through to indentured Indian labourers in Fiji, the Japanese in Hawaii or the 'South Sea Islanders' shipped to Australia in the late nineteenth century. Cultivating sugar cane was a harsh business, and it was the labour of slaves and indentured labourers that transformed sugar from a luxury item to an essential commodity. Within the space of two centuries – roughly between 1700 and 1900 – sugar became a dietary essential for all sorts of people the world over.

Clearly, there was something special, something distinctive, about sugar. People liked it and eventually came to need it. As global populations grew, especially in the nineteenth century, and as millions more expected sugar for their diet, the drive was on to satisfy their sweet cravings by cultivating sugar wherever the opportunity arose. Sugar could even be cultivated in colder climes by the late nineteenth century. The rise of sugar beet, first in Europe, then in the vast lands of North America, supplemented the world's expanding sugar production. A century later, the production of sweeteners was augmented by the development of chemical and corn sweeteners. By the end of the twentieth century, the demand for sugar was rising by about 2 per cent a year, partly to satisfy the taste of an expanding population, but also because of the rise in living standards in newly developing nations. The wider world was turning to sweet food and drink much as the West had done in the eighteenth and nineteenth centuries. As more and more people became prosperous, they demanded ever more sweetness.

From the earliest days of slave-grown sugar in the Americas, sugar was so important, so central, that it became a source of political, economic and international dispute. Even today, sugar is a topic of intense discussion between nations and within international organisations. It serves to create a confusing welter of interests, commodities and prices, all of them weaving together the various producers and consumers, different international organisations and diverse agreements into a global web spun by the world's need for sugar. What makes this even more perplexing, bizarre and even more difficult to grasp is the crucial point, now widely accepted, that sugar is actually bad for us. Indeed, medicine now affirms that sugar is bad. Full stop.

But the claims that sugar is corrupting are of very recent vintage; if it is bad today, when was it good? In many respects, sugar has been bad for centuries; it was bad for its labour force (slaves and indentured labourers), and it was bad for the ecology of sugar-growing regions. Now we learn that sugar is the prime cause of mounting ill health among nations all over the world. Nonetheless, sugar continues to be consumed in enormous volumes by more and more people. Sugar remains popular – more popular, in quantitative terms, than ever before. People still like sugar.

How, then, did all this come about? How did hundreds of millions of people come to want and to rely on sugar? If it is true that sugar is bad for us, how did the world become so corrupted by this single, simple commodity?

I

A Traditional Taste

SWEETENING FOOD AND drink has been part of human nutritional cultures for millennia. Sweetness for its own sake, sweetness to remove the bitterness of other foods and drinks, sweetness as a medical prescription, even sweetness as a religious promise – all and more have been part of human activities in countless different societies. Think, too, of the way the images and ideals of sweetness have permeated English language – the very words 'sugar', 'sweet' and 'honey' have, for centuries, represented some of life's happiest moments and the most delicious sensations. How often do people call loved ones 'sugar' or 'honey'? Most of us can recall our very first 'sweetheart'. And why, after marriage, and before embarking on a life together, do couples first enjoy a 'honeymoon'? The English language fairly abounds with the vernacular and culture of sweetness, to convey the most delicate of personal feelings – of love for another person – to the baser instincts of bribery ('a sweetener').

For many centuries, English has been replete with the language of sweetness. Middle English, for example, like the world it addressed, is littered with sweet references: to denote a loved one, a beautiful person, or someone with a good nature or disposition. Chaucer frequently uses 'sweet' to denote affection and love. So, too, three centuries later, does Shakespeare. Moreover, both men wrote in a society only marginally affected by sugar itself. The thesaurus on the very computer I used to write these words gives the following alternatives for sweetness: '*lovable, cute, charming, engaging, appealing, attractive, delightful, adorable*'.

Today, sweetness – and all that the word entails – represents many of life's great pleasures and delights. All the more curious, then, that sweetness, in the modern world, has created some of mankind's most serious personal and collective problems and dangers. Today, the desire for sweetness has become a risk to health and well-being for millions around the world.

When we think of sweetness today, we tend to think of sugar, but long before cane sugar made its seismic impact on human affairs, honey was mankind's main source of sweetness in a multitude of ancient societies. For centuries, Arabic and Persian texts, for example (in geographic, travel and cookery books), made frequent references to sweetness in contemporary cuisine and in theology. The ideal of sweetness as a delightful earthly experience – a physical sensation that is pleasurable, happy and even luxurious – is matched by the promise of sweetness as a reward in the hereafter. The afterlife is often represented as a 'sweet' experience. Nor is this merely a modern Western Christian phenomenon. In a number of faiths, heavenly pleasures come in various forms of sweetness. On earth, it took the form of honey.

Rock art from 26,000 years ago, paintings from ancient Egypt and comparable evidence from ancient Indian societies, all portrayed honey as a source of local sweetness. The world of classical antiquity similarly provides an abundance of evidence about the commonplace use of honey – as a sweetener, as medicine and as a symbol. The literature of the classical world (like English literature) is dotted with the imagery of honey. In *The Odyssey*, Homer remarked:

Never has any man passed this way in his dark vessel
and left unheard the honey-sweet music from our lips;
first he has taken his delight, then gone on his way a wiser man.
(The Odyssey, bk 12, l. 184)

Roman texts are likewise peppered with references to honey. Lucretius noted, when writing in the first century BC, that Roman doctors used honey to persuade children to swallow foul-tasting medicines:

For as with children, when the doctors try
to give them loathsome wormwood, first they smear
sweet yellow honey on the goblet's rim.
(De Rerum Natura, bk 1, l. 936)

More familiar perhaps, the Holy Bible has a profusion of images of honey. When the Lord led the Israelites out of Egypt, he led them to 'a land flowing with milk and honey' (*Exodus*, Ch. 3, v. 8). And the Old Testament, too, has numerous references to honey; the Promised Land, for instance, comes 'with milk and honey blessed'.

Honey was offered as a tribute to the gods in ancient Egypt and classical Greece, and has importance, too, in Hinduism.

Many ancient societies had time-honoured religious rituals using honey: placing a drop of honey on the lips of a newborn child; a piece of apple dipped in honey on a Jewish child's first day at school; and honey cake augurs good fortune when served on the Jewish New Year. All this is in addition to the enduring theme of honey, honey-making and bees in literature from the most ancient of recorded societies, right down to relatively modern English texts:

> *Stands the Church clock at ten to three?*
> *And is there honey still for tea?*
> (Rupert Brookes, from 'The Old
> Vicarage, Grantchester', 1916)

Honey has, then, long been both symbol and sweetener. For centuries, it was a feature of medicine and pharmacology. From ancient China and India, to classical Greece and throughout the world of Islam, honey was prescribed as medicine for a host of maladies. Like cane sugar later, honey – when mixed with other ingredients – produced medicines prescribed by Islamic and medieval doctors. To this day, it continues to be used as medication in a number of communities that have remained relatively untouched by modern medicine, and also in a variety of 'alternative' treatments which have recently found favour the world over.[1]

All this was in addition to the more obvious role played by honey as a sweetener in various cuisines. Sweet foods were (and are) especially valued in Islamic societies, partly because the Prophet liked honey and recommended it as a medicine. Even after the coming of cane sugar, sweet foods, especially desserts made with honey, have retained a special place in Islamic societies and maintain their importance in a number of ceremonies and practices.

Honey remains an important element in Islamic life. The Koran makes regular comments on sweetness: 'To enjoy sweets is a sign of faith . . .'[2] Honey was thought to be God's medicine, with a heavenly future promised in the form of rivers of honey. *The Traditional Medicine of the Prophet* (from the fourteenth century) claimed that the Prophet was fond of honey, and recommended it as a medicine for a number of ailments. Wherever Islam took root, there we find widespread and ritualised consumption of sweet foods, normally at the end of a meal, but also on specific days in the Islamic calendar. Honey was, at once, both medicine and food, its importance confirmed by a simple glance at the variety and richness of sweet foods in the Islamic diet to this day – on religious high days and holidays (the Prophet's birthday, for example), at weddings, birthdays, burials, holy days, circumcisions and family celebrations. All are marked in varying degrees by the production of lavish sweet dishes, soaked in honey and sugar. The ingredients of such desserts must, of course, conform to Islamic law.[3] Yet even *before* the rise of Islam, we know that honey had been used for a number of culinary and spiritual purposes: for nutrition, as medicine, and as a promise of future happiness.

* * *

Honey, then, had an importance and significance in a large number of ancient civilisations. It was a food in its own right and a customary ingredient in recipes and menus. But it also represented purity and morality. Both the Bible and the Koran depicted an afterlife rich in much-valued food and drink

– milk, wine and honey. Mundane earthly matters were also scattered with honey. We know, for example, of more than three hundred recipes from the eighth to ninth centuries which have come down to us in the form of *The Baghdad Cookbook* (the highest level of Perso–Islamic cuisine), although many were inherited from much earlier societies. About one third of those dishes and drinks are sweetened, such as doughnuts, fritters, pancakes, rice dishes, sherbets and other drinks.

These tastes, and the culture of Islamic cuisine and food, travelled on the back of Islam itself as it expanded throughout what is now the Middle East and the Gulf, across North Africa, into sub-Saharan Africa and into southern Europe. Naturally enough, the cultures and habits of Islamic peoples, including their cuisine and their foodstuffs, went with them. They carried with them a taste for honey *and* the recently acquired taste for cane sugar.

We know that sugar cane entered the world of Islam from India. Buddhist cuisine in India had adopted sugar as a basic ingredient as early as 260 BC and, in time, sugar began to influence the cuisine of greatly diverse societies across South-East Asia. Sugar also moved slowly westward from India into Africa, the Middle East and the Mediterranean. As Islam spread, so too did the cultivation and consumption of sugar cane. By 1400, it was being cultivated in Egypt, Syria, Jordan, North Africa, Spain and possibly Ethiopia and Zanzibar.[4]

Sugar was on the move – in all directions. In 1258, following the fall of Baghdad to the Mongols, elements of local cuisine began a protracted movement eastwards to China and to Asiatic Russia. Indeed, this global transfer was to be a feature of sugar – it was part of imperial expansion. Major empires – Greek, Roman, Islamic, Mongol, Byzantine, Ottoman and European – all absorbed foodstuffs and cuisines inherited from older empires, states and conquered peoples. And all placed great value on the sweetening powers of honey and, increasingly, of cane sugar. Sugar thus became one of the unrecognized bounties of imperial conquest and power, seized and absorbed by conquerors then carried to distant corners of the globe where it shaped new tastes and a demand for the pleasure it brought. In the European context, it was also to bring unimaginable profit.

The transformations wrought by cane sugar, however, are hard to exaggerate. Scholars agree that sugar cane originated in South Asia, but evidence for the *processing* of sugar – refining sugar from sugar cane – belongs to a much later period.[5] Over many centuries, sugar cane cultivation spread outwards from its origins. The great explosion in cane sugar production in the Americas after *c.*1600 has, however, persuaded scholars to concentrate on the westward movement of sugar, but a similar process was already at work to the east. Scholars of China, for example, have trawled Chinese sources to explain the growth of sugar cultivation and, most strikingly, the development of sugar technology and production in China. Over the long periods of the Ming dynasty (from the mid-fourteenth to the mid-seventeenth centuries) and the Qing (or Manchu) dynasty thereafter, sugar not only spread to Japan, but it also became a major commodity in Chinese trade throughout Asia, much as it did in the world of the Atlantic trading systems.

Sugar is better known when it moved westwards, along a more familiar route through Iran and Iraq, and from there to the Jordan valley, the Mediterranean coast of Syria and to Egypt, then on to other locations in the Mediterranean. Sugar cane was being cultivated in Egypt as early as the mid-eighth century and, by the eleventh century, it could be found at various points along the North African coast, on Mediterranean islands and in Spain.[6] The finished product – cane sugar – subsequently found its way, via the Crusaders, to northern Europe in the eleventh century. Of course, sugar was only one of a number of foods that were transplanted westward in these years, travelling and settling like the botanical flotsam and jetsam of human and religious migration and upheavals. Rice, cotton, eggplants, watermelons, bananas, oranges and lemons travelled along similar routes.[7]

Not surprisingly then, there is an abundance of evidence about sugar in early Arabic literature, with detailed accounts of sugar, its pleasures and its alleged benefits. It crops up in all sorts of literary sources. In *The Thousand and One Nights* (parts of which date back to the ninth century), a conversation between a poet and a slave described sugar cane thus:

> It is shaped like a spear, but has no head.
> Everyone loves it.
> We often chew on it after sunset during Ramadan.[8]

These and similar passing references reveal a remarkable feature about the story of sugar – that from its earliest days down to the present, sugar has attracted a great deal of contemporary attention. The spread of Islam involved not merely conquest and conversion of enormous swathes of land and people, but the scattering of cultural habits, ranging from the world of print and learning to modern science, medicine and cuisine, with a number of scholars describing the spread of sugar production and consumption. We learn, for example, about sugar in the tenth century from an Arab geographer. In 1154, it was the turn of a merchant, describing his travels, to provide a description of sugar cultivation and production. We also have descriptions of sugar processing and the financing of sugar from late medieval Egypt.[9]

The spread of sugar around the Mediterranean was not simply a matter of cultivation, but involved new systems of agricultural production, methods of irrigation and technology of sugar processing, all in addition to the financial ability to develop sugar production and to distribute the final product – cane sugar. By the time of the Islamic defeat and expulsion from Europe in 1492, well-developed and well-known patterns for sugar production had been established. It was to be used (though transformed out of all recognition) by Europeans when they explored and settled the Atlantic islands and, later still, the tropical lands of the Americas.

The first major English encounter with sugar was in Palestine during the First Crusade of 1095–1099. Sugar cane saved Crusaders in times of starvation, and nurtured a taste for sweetness (and for other exotic commodities) which survivors took home with them. But sugar was both rare and costly, and was naturally restricted to contemporary elites. We can catch a glimpse of it in medieval household accounts documenting the purchase and storage of foodstuffs in the larders and kitchens of palaces, castles and religious houses. Monks in Durham described their sugar as '*Marrokes*' and '*Babilon*'. The Earl of Derby's sugar was listed as '*Candy*' (the contemporary name for Crete), while other recipe books described sugar as '*Cypre*' (Cyprus) and '*Alysaunder*'. The Great Wardrobe accounts of Edward I for 1287 recorded the purchase of 667lb of sugar, 300lb of 'violet sugar' and a huge 1,900lb of 'rose sugar' (the last two, sugar mixed with powdered flower petals, were used as medicines).[10] All of these sugars were clearly from the

Mediterranean and had arrived in England via merchants in Venice and Genoa who, in their turn, had acquired sugar from producers scattered around the Mediterranean.

The volumes of sugar produced were small, but they increased as the taste and fashion for sugar spread among Europe's prosperous elites. By the thirteenth century, sugar was regularly used in elite English households.[11] In 1319, for instance, Nicoletto Basadona carried 100,000lb of sugar and 1,000lb of 'candy sugar' to England.[12] Sugar was also common in French cuisine by the fourteenth century. In that same century, records show increasing amounts of sugar imported into the Kent port of Sandwich. (Imports into southern ports may explain why sugar took hold initially in the south, and came more slowly to the north of England.) Sugar was also landed in Boston, Lincolnshire, from Amsterdam, Calais and Rotterdam, and it was also imported through the more distant ports of Devon. By the late sixteenth century, 'comfit makers' – specialists in making confectionery from sugar – began to appear in major English provincial towns.[13]

By the sixteenth century, sugar was widespread in England; the Earl of Northumberland's 'clerk controller' ordered more than 2000lb of sugar for his Lordship's kitchen.[14] Sugary confections – fruits in sugar, sweet cakes, preserved sweetened fruits – all had become so prominent a feature of royal households that monarchs appointed officials specifically to take charge of the confectionary department. The same official became skilled in the preparation of various forms of sugar and of sweet foodstuffs for the royal table. In time, in the larger royal households (notably in Hampton Court), there was a special bakery for sweetened foodstuffs. Formal recipes and menus prepared for royal and aristocratic households now included include sugary desserts.[15] Servants were taught when and how to use sugar in the course of a meal (to be offered in a sauce for partridge and pheasant, or to be sprinkled on baked herring, for example). Revealingly, new eating utensils were designed to eat sugary concoctions. Elaborate and costly plates were reserved for sweetmeats, and special forks were provided to lift and raise sticky sweets to the lips. From the 1580s, early English recipe books described how best to use sugar, such as for stuffing rabbits and preserving fruits.[16]

European elites of all sorts – royals, aristocrats and clerics – adopted sugar both as a feature of their elaborate cuisine and as a means of flaunting their status and rank via ornate models and statuettes modelled from sugar. In this they were copying an older Islamic tradition of using sugar in elaborate displays of power and wealth. There were plenty of ancient tales of rulers and sultans organising elaborately sculptured displays to celebrate religious festivals. One visitor to Egypt in 1040 reported that the Sultan had used 73,000kg of sugar for a display which included a tree made of sugar. Another account from 1412 told of a mosque built of sugar – all of it consumed by beggars when the festivities ended.[17] At an Ottoman festival in Istanbul in 1582, hundreds of sugar models were created to celebrate the circumcision of a son of the Sultan. It had models of animals and a castle which were so heavy it required four men to carry them.[18] Among other things, such elaborate displays of sugar revealed the costliness of sugar; only those with very deep pockets could afford to finance such ornate displays of sugary confection. Coronations, military victories, sacred festivals – all and more were marked by elaborate sculptures in sugar.

As sugar spread from the Mediterranean to mainland Europe, largely via Venice, so, too, did the fashion for elaborate sculptured sugar. Europeans chefs, cooks and bakers adopted the ingredients and habits of Arab societies, but adapted them to local needs and tastes using moulds, or worked from a sugary paste. Chefs and their assistants quickly acquired the necessary skills; they created lavish and ornate sugary displays for festivals and ceremonies among Europe's elites from the thirteenth to the seventeenth centuries.

Led by royalty, the French were both the European pioneers and the perfectionists of this new culinary art. Guillaume Tirel (nicknamed Taillevent), who worked in French royal households between 1326 and 1395, left behind a manuscript of recipes which made frequent and varied use of sugar for royal dishes.[19] Although honey continued to be used, from the thirteenth century onwards, costly imported sugar cones became increasingly common in wealthy households. Often, though, that sugar was crude and had to be refined and clarified again in the local kitchen before being prepared for consumption. It remained far beyond the pockets of all but the wealthiest and most privileged in society – although we know it was for sale in a London grocer's shop in 1379.[20]

By the sixteenth century, French cuisine was using sugar for three main purposes: to sweeten dishes; to preserve fruits, flowers and vegetables; and to mould into decorative ornaments and models, or to glaze. Sugar was mixed with various gums and pastes – notably with almonds to produce marzipan – to make a dough that has remained a basic ingredient of confectionery up to the present day. At much the same time, French cookbooks outlined the methods needed for boiling sugar to produce various syrups and crystallised sugary items (compotes, barley sugar and caramels).[21]

Best remembered, however, were the lavish sugary sculptures. In 1571, the city of Paris organised an elaborate dinner for Elizabeth of Austria, the new Queen of Charles IX. All who saw the event agreed that it was the most elaborate ever seen. Each course was heralded by trumpets and each was based around an appropriate theme. Dinner was followed by dancing, which was followed by a 'collation' – preserves, sugared nuts, fruit pastes, marzipans, biscuits and a variety of meat and fish – all of them fashioned from sugar paste. The main dining table was decorated with six large sugar sculptures telling the story of how Minerva brought peace to Athens.[22]

Sugar had also become a central item in the way the dining table should be set out, particularly for the most formal of meals. Sugar sculptures took their place alongside floral displays and elaborate silverware on the most important dining tables. The table arrangers even copied the examples of contemporary landscape designers to create elaborate landscapes on the tables, and all were scattered with sugar moulds and sculptures. Skilled confectioners used sugar (in a host of colours) and marzipan to create whatever scenes and images their masters and mistresses demanded.[23]

Such displays of power, wealth and status were important, and chefs in palaces and stately homes perfected the art of fashioning sugary blends into edible sculptures to astound, impress and feed. Mixing sugar with nuts and gums, or pouring liquid sugar into moulds made specifically for the purpose, cooks vied to outdo each other in their elaborate concoctions, to grace the tables and attract the admiration of guests at formal banquets and state ceremonies. Known as '*soteltes*' (subtleties), they were designed initially to be consumed *between* courses, and might be shaped in the form of fish or meats. In time, however, they took on great significance in conveying messages from rulers to their rivals, friends and enemies, and were designed to impress guests by flaunting the host's status and wealth.

Other privileged members of society soon adopted the sugary habits of their rulers. High-ranking clerics and prominent academics, for example, all found sugar sculptures a perfect reflection of their own status and positions. When Thomas Wolsey was installed as Cardinal in Westminster Abbey in 1515, he ordered an extravagant display of churches, castles, beasts, birds – and a chess set – all made from sugar.[24] For his installation in 1503, the Vice Chancellor of the University of Oxford ordered a display of 'the eight towers of the university', its officers and the King – all made from sugar.[25] In 1526, Henry VIII employed seven cooks to devise an elaborate sugar banquet at Greenwich which displayed a dungeon and a manor festooned with swans and cygnets, while another chef created a tower and a chessboard, all 'garnished with fine gold'.[26] More daring still were sugary displays of genitalia crafted for the amusement of dinner guests, though more formal religious or diplomatic dinners were graced by more tasteful sugar sculptures, such as religious or royal images to fit the occasion.[27] It is no surprise to learn that the French and English courts suffered dreadfully from dental problems – rotten and missing teeth, gum disease, collapsed mouths and disfigured appearances. All were a consequence of sugar.

As wealth spread to a new merchant and trading class (many grown fat on the pickings of overseas and imperial trade and settlement), so, too, did the luxurious habits of their betters; they also began to use sugar to impress and entertain. In common with other luxury items, however, the more popular its base, the less potent the message, and as sugar became more widespread and cheaper by the late sixteenth century (courtesy of African slaves in the Americas), the prestige of elaborate displays of sugar lost their effectiveness. English elites tended to buy their sugar in London but, by the mid-seventeenth century, sugar was available in the smallest of provincial towns – in Mansfield in 1635 and Rochdale in 1649, for example. In Tarpoley, Cheshire, in 1683, locals could buy sugar from Ralph Edge, the local ironmonger.[28] By the time sugar entered the homes of humbler sorts, it had lost its social cachet among the wealthy.

The commonplace use of sugar in household affairs was reflected in early cookery and recipe books. Distinctively English recipe books first appeared in the 1580s, and included sugar as an ingredient, both for preserving fruit and for cooking. Gervase Markham's manual, *The English Housewife* (first published in 1616 but drawing on much older advice and recipes), is strewn with recommendations for the use of sugar in cooking and food preparation. Sugar was thought ideal, for example, in salads, pancakes, veal roasts, fritters, to enhance liver, for a number of sauces, for oyster pie, for a string of puddings, pies and jellies, for spice cakes and, of course, for 'a sugar plate'.[29] This handbook also thought that the ideal housewife should not restrict herself solely to cooking. She was also charged with the household's health and well-being, and the book offered instructions about contemporary nursing and healthcare, and for any ailment or accident that might happen. Even here, sugar was invaluable, and it was recommended in a cordial 'for any infection at the heart' and for 'a new cough' and for 'an old cough'. Sugar was recommended for an eye problem, for consumption and to staunch the blood, 'for the wind colic', and even 'for any old sore'.[30] Sugar was now as medicinal as it was tasty, and as practical as it was symbolic. If it could impress in sculptured displays, it might, if administered properly, even provide succour to the sick and cure the infirm.

Sugar had taken its place in the kitchen not simply as an ingredient, but as a medicine, and the explanation is again to be found in the spread of Islam. The development of a new Islamic orthodoxy saw the emergence of a new kind of Islamic medicine, much of it rooted in pronouncements of the Prophet and his followers, but also encouraged by the rise of new learning, centred on Baghdad, and on the translation of ancient, classical texts – such as the Greek medical writings of Galen – into Arabic. Hence Galen's medical ideas permeated the world of Islam – and beyond. A rich medical literature emerged, most notably the dominant compendia which provided a digest of medical learning, along with questions and answers for anyone interested in medical issues.

Islam also spawned a new breed of doctors whose work and research, now available in printed form, advanced learning and understanding of the human body, its ailments and treatments.[31] The most famous and influential, Al-Razi (865–925), recommended that 'unpleasant tasting drugs should be made palatable'.[32] Like the Greeks and Romans before them, Islamic physicians, and those who followed (especially Spanish and Jewish authorities), found sugar and honey ideal antidotes to the bitterness of certain medicines. It was a slow, gradual process, but sugar became part of Islamic and then European pharmacology.

Medicine had also been helped by the vast geography of Islam, by yielding an astonishing array of flora and fauna, and of minerals and animals to be used as medicines. By the thirteenth century, pharmacists had lists of upwards of 3,000 items used in medicines, many of them exotic items culled from distant tropical regions. Sugar was only one of a long list of such items, but it quickly found a special niche, both for its own sake, and more broadly for the way it made unpleasant drugs palatable.

These Islamic medical and pharmaceutical traditions spread to Western Europe. Apothecaries (from the word *apotheca*, meaning a place where wine, spice and herbs were stored), dispensed sugar alongside, or mixed with, other medicines. Robert de Montpellier, 'spicer-apothecary' to King Henry III, opened London's first pharmacy in 1245 and, among his wares, he offered 'electuaries' – mixtures of spices and herbs bound together by sugar and prescribed for the sick. At the end of his life, Henry VII was treated with sugar mixed with rose water, violets and cinnamon.[33]

In France, Louis XIV employed Monsieur Pomet as his 'Chief Druggist'; Pomet later published *A Complete History of Drugs*, a work translated and published (edited and added to) in London in 1748. It devoted five pages to sugar – its nature, cultivation, and culinary and medical use. Quite apart from all the tasty sweets, desserts and drinks provided by sugar, it was, so the author claimed, good for the breast and the lungs, for asthma, coughs, for kidneys and the bladder. However (and here Pomet must have looked closely at Louis XIV himself), 'It rots and decays the teeth . . .' Having listed all the places where sugar was grown, the book claimed that 'now our fine Jamaica and Barbados Sugar is inferior to none . . . and next to them is it reckon'd the Lisbon sugar . . .'[34]

By around 1600, sugar had undergone a remarkable transformation. What had, up to this point, been the preserve of the rich and powerful, was now available in the humblest of shops and in the smallest of villages. Tarpoley was a long way from the French royal court; even further, in time and distance, from the great centres of Islamic learning and medical science. Yet there was a link – a progression – from one to the other. Once the monopoly of kings, by the mid-seventeenth century sugar could be bought from a humble ironmonger in the north of England. It had begun to change from an expensive luxury to the everyday necessity of ordinary people. Even more curious is the fact that this massive change in direction and fortunes was all made possible by the brutal exploitation of vast numbers of African slaves. Tons of sugar now found their way to the docksides of Europe, and from there to local refineries and, finally, onwards to markets, fairs, shops and travelling salesmen, reaching consumers across Western Europe and, later, across the globe. Sugar was to become an everyday item in the shops and shopping baskets of humble people.

2

The March of Decay

B Y THE REIGN of Elizabeth I, sugar was hugely popular
among the upper echelons of English society. They ate
and drank it in abundance (Shakespeare's Falstaff loved his
sweet wines made even sweeter by the addition of sugar), and
they revelled in lavish, sugary displays of power and influence.
When the Queen progressed through Hampshire in 1591, the
Earl of Hertford laid on a firework display followed by a
banquet dominated by 'Her Majesty's arms in sugar works . . .
Castles, forts, ordnance, drummers, trumpeters and soldiers of
all sorts, in sugar works . . .' Exotic beasts and birds, snakes,
whales, dolphins and fish – all made from sugar – were paraded
for the Queen's pleasure. The monarch had a very sweet tooth.
In 1597, the French Ambassador wrote of the 64-year-old
monarch: 'Her teeth are very yellow and unequal . . . Many of
them are missing so that one cannot understand her easily
when she speaks quickly.' A year later, a different visitor thought
her teeth were black. Even by the late sixteenth century, it was

already clear that sugar wrought great damage to people's teeth.[1] Today, when faced with dental problems that can be quickly and painlessly solved, we tend to wince when thinking of our ancestors' dental sufferings. In fact, rotten teeth and painful dental treatment are relatively modern phenomena – and are, overwhelmingly, associated with the history of sugar. We have become aware of this partly because of the advances in modern science, and we're painfully aware now that tooth decay is particularly destructive when sugar reacts with bacteria to produce an acid that attacks the enamel. But, in recent years, the associated work of archaeologists has revealed that our forebears did *not* suffer the widespread dental problems we often imagine – not until the coming of refined sugar. Curiously, the devastating explosion of Mount Vesuvius provides some useful clues.

On 24 August AD 79, Mount Vesuvius erupted in what, alongside Krakatoa in 1883, proved to be perhaps the most famous volcanic eruption in human history. It destroyed the towns of Pompeii and Herculaneum, and much else in the close vicinity, killing untold thousands via waves of roaring, roasting heat which were followed by inundations of ash and volcanic lava. The ash which engulfed the towns and their inhabitants eventually hardened into pumice. In time, the bodies trapped in those shells of pumice rotted, leaving behind mere skeletons. Modern archaeologists using new technology and materials have created casts of the victims by pouring plaster into the shells. Teams of scientists, archaeologists, radiologists, doctors and dentists have recently analysed these human remains, and subjected them to experiments which would have been impossible a mere generation ago. Those human remains, which have been buried for centuries under layers of volcanic ash and lava, have begun to yield evidence about the condition and health of those who died. The remains of thirty people, examined by modern CT scans, have revealed, among other things, remarkably good dental condition. Scans, X-rays and dental analysis suggest that the victims (men, women and children) had no real need of dental treatment; few of them had cavities. When they died, their teeth were in very good condition.[2]

We know a great deal about their diet from a variety of historical sources. It was a fibre-rich Mediterranean diet, characterised by lots of fruit and vegetables. Most crucially, perhaps, we also know that they enjoyed a sugar-free, or very low-sugar, diet. Theirs was a balanced diet, very similar to the one proposed by modern medical dieticians seeking a healthy alternative to modern sugar-soaked, fat-laden foodstuffs. Above all, the victims of Vesuvius did not eat refined sugar, and the teeth of those killed by Vesuvius in AD 79 provide a vivid example of what teeth look like *without* the consumption of sugar.

The Vesuvius example is eye-catching but not exceptional. A number of archaeological and medical examinations of teeth from a range of ancient burial sites tell a very similar story. Almost 1,000 British examples, taken from sites ranging between the Iron Age and the late medieval period – some 2,000 years – showed no deterioration in their dental condition. Specific case studies confirm the pattern; an examination of 504 Anglo-Saxon examples showed none of the kind of tooth cavities caused by sugar.

The pattern begins to change, however, in the seventeenth century. And by the nineteenth century – the years of Britain's urban and industrial transformation – burial grounds tell a very different story. The Victorian dead offer up repetitive evidence of widespread dental problems – bad teeth, numerous cavities, decay and poor overall dental health. What lies behind this remarkable transformation is the story of sugar.[3] We have dental data in Britain which spans more than 2,000 years. And there are even more sweeping studies of *global* dental and archaeological data that reinforces the pattern. Time and again, the main or sole cause of dental decay was the natural process of ageing. In the South Pacific, in ancient Egypt and among native peoples in North America, dental decay was a function of age – not diet. The overall conclusion, drawn by a renowned professor of dental surgery, was that 'prior to the seventeenth century (and much later in rural areas), people probably did not suffer as greatly from dental pain as might be imagined'. And that was because they did not eat or drink sugar.[4]

Compare this to the evidence of widespread dental problems in the current British population. Extensive dental decay – most worrying and most widespread among the young – is a regular and commonplace topic for discussion among medical experts, and thus in the media. Of course, the precise mix of dietary factors which produces the problems of dental ill health in Britain (and elsewhere) are complex, but today the only doubts about the role of sugar in creating those problem are raised by the sugar and food industries and their paid lobbyists. Not surprisingly, when such stories surface in the popular press about, say, the generally good dental health of ancient Romans, they do so in the form of an eye-catching headline: 'ANCIENT ROMANS IN POMPEII HAD "PERFECT TEETH".'[5] The beginnings of widespread modern dental problems were first noticed among the wealthy, not surprisingly because they were people who could afford lavish helpings of sugar. Though Elizabeth I's problems (and those of her prosperous subjects) were conspicuous, they paled when compared to the dental difficulties of French royalty and aristocracy. Their dental problems can even be seen today in the form of French portraiture.

In 1701, Hyacinthe Rigaud painted a lavish formal portrait of the 63-year-old King Louis XIV, the 'Sun King'. It is a dazzling, sumptuous display of royal power, draped in all the appropriate symbols of wealth and regal authority. This small, bald man actually looks tall, his head engulfed by a huge wig of curly hair. The artist's skills and artifice, however, could do nothing with the King's mouth and cheeks. Louis was 'a ruler with not a tooth in his head'. Louis XIV had lost all his teeth by the age of forty, despite a personal retinue of medical staff providing the best of contemporary medical and dental care. Though they monitored the King's general well-being, they paid no attention to his consumption of sugar. Neither was Louis alone in his toothlessness; dental decay and tooth loss were common throughout Louis' glittering court, much more so indeed than among his poorer subjects. Acute observers noted that the poorest of street urchins – i.e. those with no access to costly, sugary foodstuffs and luxuries – seemed to have better teeth than their social superiors. In the words of Colin Jones, 'Empty mouths were a fact of adult life of Europe's Old Regime of Teeth . . .'[6] As sugar spread lower down the French social scale, so, too, did dental problems.

The French mixed sugar with their chocolate, coffee, tea and lemonade, and their cooks heaped it in profusion into their foods and delicacies. In polite and courtly circles, drinking and eating these sugary concoctions became rituals of polite and fashionable behaviour. Social fashion compounded personal taste and dictated the increasing enjoyment of sugar. All this (plus the growing addiction to tobacco) created widespread dental problems among the French. Again, the evidence is stark, and not merely in the portraits of the rich and powerful. Archaeologists working in French graveyards have found very different evidence from modern researchers at Pompeii; French teeth in the seventeenth and eighteenth centuries were rotten.[7] The dental problems of France's ruling elites remain most vividly on display to this day in contemporary portraits. Like Rigaud's picture of Louis XIV, any number of later paintings of the rich and famous rarely showed the sitter's teeth, largely because they were so sparse or rotten, although it is also true that contemporary artistic conventions dictated strict facial postures and mannerisms. Smiling was a sign of vulgarity and low life, laughing even more so, an indication of those passions that were best kept from public view.

In the course of the eighteenth century, however, the emergence of a new sensibility led to a marked stylistic change, and a growing acceptance that smiling – even the revelation of white teeth – was an acceptable social trait, both in public and on canvas. For that to happen, however, the skills of a new breed of dentists were required, men able to preserve teeth, to heighten their whiteness – men who could enable their patients to smile in public. Such skills – even in their basic eighteenth-century form – were beyond the reach of all but the richest members of society. Even then, there were limits to what could be achieved. They were struggling against the rising consumption of sugar.

Sugar poured into France from its booming colonies in the Caribbean – from Guadeloupe and Martinique but, above all, from St Domingue (later named Haiti). Like all the major European maritime and colonial powers, France had developed important outposts in the Americas. Although not the pioneers, the French had rapidly caught up with the Portuguese, Spaniards, Dutch and British in the Americas. It was accepted that the key to rising French colonial fortunes lay in India and the Americas, and French rivalry with the British for global ascendancy would be played out in both those regions. In the Americas, but especially in the Caribbean islands, the critical ingredient for local success was labour provided by African slaves, with all the complexities of shipping enslaved Africans across the Atlantic. In the islands, enslaved Africans laboured to produce a variety of tropical commodities, with sugar dominating.

In the early eighteenth century, Jamaica was the Caribbean's main sugar producer but, by 1770, France's major Caribbean colony had become the world's largest exporter of sugar. By then, St Domingue produced 60,000 tons a year (compared to Jamaica's 36,000 tons) and all was made possible by massive importations of Africans. By the time slavery in St Domingue – and the economy which it sustained – were destroyed by the great slave upheaval after 1791, no fewer than 790,000 Africans had been imported into St Domingue. The sugar and, at higher altitudes, coffee produced by the Africans was the source of both French wealth and French dental troubles. It was the cause and occasion of the sunken, hollow cheeks, the slack jaws, the toothless heads and the rotten teeth. It was as if the enslaved were getting their revenge for the abominations heaped upon them in the Caribbean.[8] Thanks to events 5,000 miles away, sweet coffee became a national French drink, most strikingly in the cities. As the volumes of slave-grown commodities increased, the cost of sweet coffee fell. In Paris, cafés multiplied and coffee-drinking established itself as part of public and private social life (the word 'café' is itself a giveaway to the entire story).

But coffee (like tea) was a naturally bitter drink, and Europeans (and Americans) added sugar as an antidote to the bitterness; they liked their coffee sweet. One English visitor thought that the French heaped so much sugar into their coffee that the spoon could stand up in the cup. Sweet coffee was everywhere, served in royal palaces and sold by Parisian street vendors working from a wooden bench. By the late eighteenth century, an enormous variety of Parisian coffee shops catered for all sorts and conditions, from the costly and fashionable down to the lowest of dives offering a cup of hot, sugary coffee and shelter for the poor. Female street peddlers sold a cheap drink made from coffee dregs mixed with warm milk to people setting out for work in the early morning.[9] But everywhere, in palaces and on the streets, the accompanying sugar burrowed into, and rotted, the teeth and gums of the coffee-drinkers.

It is no surprise then that French portraits failed to show the sitters' teeth. Whatever the contemporary convention about facial appearances – the fear of being considered unbecoming (especially for men) – smiles were more likely to reveal a miserable dental condition. The modern (especially the American) ideal of lavish displays of rows of sparkling teeth (sometimes more akin to piano keys than teeth) had to await the rise of modern and costly dentistry and the skills of the orthodontist. Cost permitting, late twentieth-century dentistry and medicine can make good whatever ravages diet had inflicted on the teeth of earlier generations.

* * *

It is an enormous irony (and scandal) that, in 2015, a study of the dental health of modern British children revealed a fact that would have been familiar to French aristocrats three centuries earlier. Large numbers of those children do not like to laugh or smile because of the poor condition of their teeth. More than one third of twelve-year-olds, and more than a quarter of fifteen-year-olds, claimed to be embarrassed when laughing or smiling because of their teeth. Of course, this simple fact masks a complexity of issues, but it caught the eye of the British media. Sensational headlines conveyed a simple message. *The Times* thundered: 'ROTTEN TEETH ARE SECRET REASON WHY TEENS DON'T SMILE'.[10] Louis XIV would have felt at home.

Yet the teeth of British children today *ought* to be excellent. Today's children are born into a wealthy society which aims to provide healthcare from the cradle to the grave. Their health and well-being are monitored from the moment they are born. Moreover, the National Health Service (despite its recurring difficulties) offers this lifetime range of healthcare free. For all that, it has become clear that the teeth of large numbers of British children are in a poor state. Perhaps understandably, the teeth of children from poorer homes are worse than those from more prosperous backgrounds. Even so, the long-term trends are alarming.

Starting in 1973, a survey was launched of the dental health of children in England, Wales and Northern Ireland, and has been repeated every ten years. It has shown that upwards of one fifth of the children reported having trouble eating – because of dental problems. Almost a quarter of the parents involved had to take time off work to attend to their children's dental care. One child in seven between the ages of five and fifteen suffered from severe or extensive decay. At its most extreme, these dental problems arrive as medical emergencies at the nearest hospital. In 2011–2014, almost 26,000 children, aged five to nine, were admitted to hospital in England for extensive tooth extraction under general anaesthetic, an increase of 14 per cent from 2011. Clearly, this was no longer a simple dental matter; the Royal College of Surgeons expressed alarm both at the problem itself (not least because most of it was preventable) and because of the strain placed on hospitals.[11]

Despite the NHS, and despite the extensive and costly publicity about dental hygiene, large numbers of British children suffer from alarming dental problems. The medical staff most closely involved in monitoring and dealing with the problem are in no doubt about the cause. Dr Sandra White, England's Director of Dental Health for Public Health England in 2016, was blunt. She stated that although tooth decay among the young was slowing down, there remained an urgent need to reduce the amount of sugar in children's diets.[12] It is an irony of enormous proportions that modern British children find themselves confronting the same problem that beset Louis XIV – too much sugar.

3

Sugar and Slavery

CANE SUGAR SPREAD from plantations in the Mediterranean into northern Europe largely via major trading dynasties in Catalonia, Genoa and Venice. Compared to what followed, the volumes of sugar traded to Germany, the Low Countries and England were tiny. Sugar was a small-scale luxury trade, but it proved astonishingly influential. It created a taste for sweetness, and a commercial and financial system to satisfy that taste. It was also clear that the cultivation of sugar cane, and new systems of processing the cane in the Mediterranean (using rollers to crush more juice from the cane), provided the basis for a lucrative commercial venture, which was largely the work of the financial and commercial acumen of Italian merchants and financiers. All this was in place *before* Europeans began to trade and settle outside the Mediterranean, on the Atlantic islands, and long before the settlement in the Americas.

The labour used on early Mediterranean sugar plantations had been a mix of free and slave labour. Slaves were procured from warfare on the edges of Europe, notably between Christians and Muslims, and from conquests in North Africa, but sugar plantations also involved costly investment, with the money provided by Spanish and Italian city merchants. The resulting crude sugar often needed further refinement, initially in Antwerp, but later in many of Europe's major ports.

The broad outlines of the modern sugar economy were thus in place long before Europeans embarked on their American adventures. Sugar was cultivated and processed on plantations financed by European merchants and bankers, and used slaves and free labour working side by side. The end product – cane sugar – was then refined in northern European cities before being marketed and sold to wealthy consumers across Europe. In time, this simple sugar economy was to become a mass market, and sugar itself was transformed.

That transformation began when Europeans embarked on their remarkable age of maritime exploration in the fifteenth century. Led by the Portuguese, Europeans began to edge their way into the Atlantic, then south along the African coastline (initially in search of gold and spices, ultimately seeking sea routes to Asia). En route, Spain and Portugal settled the Atlantic islands – Madeira and the Canaries first, then São Tomé and Príncipe, both much further south in the Gulf of Guinea. Sugar cultivation was an obvious line of commercial development and land settlement.

In 1425, Henry the Navigator equipped the early colonists to Madeira with sugar cane plants from Sicily. By the end of that century, the island was producing substantial volumes of sugar. Meanwhile, Spanish settlers overcame the resistance of indigenous people to take possession of the Canaries, and they, too, planted sugar cane. Throughout these Atlantic islands

– the Azores, the Canaries, Cape Verde Islands, Madeira, São Tomé and Príncipe – the conquering European settlers planted sugar cane where it was possible. In some places, they discovered that other crops were more suitable and profitable (wine and wheat in the Azores, for example). But sugar soon proved its worth on Madeira, an uninhabited island which attracted Portuguese settlers to the prospect of land, and which they turned over to wheat and sugar cultivation (with help from Genoese and Jewish financiers). Sugar became the island's most lucrative commodity and, by the end of the fifteenth century, Madeira was the West's largest sugar producer. It also employed slave labour from Africa, as well as from the Canary Islands. Most sugar cultivation was in the hands of small cane farmers who sent their cane to be processed in the nearest factory, some of which used the latest water-powered technology. By 1470, Madeira was producing 20,000 *arrobas* (230 metric tonnes) of sugar; by 1500–1510 (the peak years of local production) that volume had risen to 230,000 *arrobas* (2,645 metric tonnes) of various kinds of sugar.[1]

This pattern of sugar cultivation in the Canaries was to become familiar in the Americas (though on a vastly different scale). The sugar was cultivated by both male and female slaves – Africans, people of mixed race and slaves from the Canaries (although that was outlawed at the end of the century) – and all this took place on small-scale plantations or '*engenhos*'. Compared to what was to follow in the Americas, this seems miniscule but, in essence, it formed a blueprint that sugar planters in the Americas were to follow later.

By the early sixteenth century, sugar was considered the most important enterprise on Tenerife. A good local plantation produced about 50 tons of sugar a year and, in a fruitful year, the island produced about 1,000 tons of sugar. The labour was enslaved, and overwhelmingly male; in time, it was increasingly African, with the managerial, technical or supervisory staff made up of free Portuguese. There was also a scattering of poorer sugar cane producers who simply grew sugar cane and dispatched their cane to the nearest factory for processing. Once again, the finance for all this came from Spanish and Italian financiers, with the finished product dispatched to mainland Europe via major European commercial houses.

Throughout the sixteenth century, sugar from the Canaries was an important export to Europe (and remained as important as the sugar exported from the Spanish Caribbean until c.1600). Eventually, however, and under competition from sugar from the Caribbean, sugar in the Canaries gave way to wine production. The Cape Verde islands, also thought a likely sugar producer, were 400 miles off the African coast and had a dryer climate, and settlers were reluctant to move to so remote a place. The initial sugar industry never really took off. What finally undermined the islands' fledgling sugar industry was competition elsewhere, and the fact that the Cape Verde islands became an important way station in the Atlantic slave trade; it had other commercial temptations to offer the enterprising settlers and agriculturists.

On the eve of the settlement of the Americas, the widespread cultivation and processing of sugar was perfected in what might seem, at first, an unlikely location: two small islands lying close to the African coast in the Gulf of Guinea. The Portuguese landed on São Tomé in 1471, one of their discoveries as they made their uncertain way south along the African coast. It was uninhabited and ideally suited to settlement. The cultivation which developed followed the pattern established in Madeira and the Azores. Sugar cultivation started at once, with the help of settlers already familiar with sugar production, and with finance, again, from Italy. By the mid-sixteenth century, the sugar economy of São Tomé was booming, yielding 150,000 *arrobas* of sugar. At its peak, 200 sugar mills dotted the landscape and the island's population rose to 100,000. More striking still, the labour force was increasingly African – and enslaved.[2]

Enslaved Africans had previously passed through the islands along early Portuguese slaving routes which saw Africans shipped from one coastal African society to another. This early European slave trade involved selling Africans to other Africans. But São Tomé lay only 320km off the coast of Africa, and African slaves were readily available for islanders in return for the variety of goods offered by European traders. From the first days of settlement, São Tomé had been an entrepôt for the movement of goods on the routes of trade and exploration along the African coast. Now it became a destination for coffles (groups of chained slaves) of enslaved Africans destined for São Tomé's sugar fields.[3] The numbers involved (compared to what was to follow) were relatively small; in 1519, more than 4,000 Africans were delivered to the island. A few years later, the Portuguese Crown was obliged to regulate the slave trade to the island. By the middle of that century, there were some 2,000 African slaves working in the island's sugar fields, but perhaps three times that number were waiting in pens to be transported elsewhere.

It was so easy – and cheap – to acquire African slaves that they were used lavishly on the island's sugar plantations. Some local properties employed large communities of Africans, upwards of 150. They came from widely scattered regions on the African coast – from Benin, Angola and Senegambia – and were subjected to an intensive labouring regime which left little free time (and some of them were assigned to cultivating foodstuffs for slave labourers). Here, in embryo, was a system that would have been recognized by observers of the Caribbean sugar economy three centuries later.[4]

São Tomé was then a thriving sugar economy in the sixteenth century, its sugars destined for the refineries of Antwerp and from there to fashionable tables across Europe. But it was an industry which crashed almost as quickly as it arose, although not because of a collapse in the demand for sugar. The astonishing European growth in demand for sugar in the following centuries was satisfied by planters not close to the African coast, but thousands of miles away on the far side of the Atlantic. Brazil led the way when settlers, tempted by the apparently boundless plenty afforded by Brazilian land, migrated from São Tomé bringing their African slaves with them. By 1700, São Tomé's sugar industry had virtually disappeared – the island even imported sugar from Brazil.

* * *

In a protracted journey westwards, sugar cane first crossed the Atlantic with Columbus on his second voyage in 1493. Columbus, who had lived in Madeira and worked for a Genoese company which dealt with sugar, was alert to the commercial prospects afforded by sugar. Equally, all European adventurers and pioneers in the age of discovery were looking for whatever commercial and agricultural opportunities were afforded by distant lands. They were seeking land for settlement, for

34

agricultural experimentation and development, and they were keen to shift seeds, bulbs, plants and cuttings from one corner of the globe to another. Few seemed to yield so remarkable a bounty as sugar cane.

Introducing sugar cane into the tropical Americas seemed an obvious thing to do. It had, after all, proved its worth in a number of different locations around the Mediterranean and more recently in the Atlantic. It yielded good returns to European investors, had been the cause and occasion of agricultural and proto-industrial development, and it had proved that it thrived well with African slave labour. It was, however, one thing to shift Africans from their homelands to São Tomé, quite another to transport them clean across the Atlantic. The key to the entire story, however, is that sugar had found a thriving market in Europe itself. The sugar pioneers in the Americas knew that Europeans were eager for ever more sugar. They also knew that Africans were available to work in their fields.

The first efforts at sugar cultivation by the Spaniards in Santo Domingo, Cuba and Puerto Rico produced only modest returns, mainly because Spain was interested in more lucrative corners of their empire in the Americas. Who would prefer the hard slog of sugar planting when the fabled riches of El Dorado lay on the mainland, just across the waters of the Caribbean?

The same could not be said of Brazil, though. There, the initial commercial attraction was timber, although from the first it was clear that settlers and their Portuguese masters were interested in experimenting with sugar. The first small cargoes of Brazilian sugar were sold in Antwerp in 1519, but that began to change substantially during the 1530s when experienced planters and financial backers, equipped with sugar cane, took root in newly developed Portuguese settlements. The impetus was Portuguese imperial backing, in the form of franchises – 'captaincies' – given to men to settle and develop specific regions of Brazil. Sugar was transplanted from Madeira and São Tomé and, although many settlements failed, those that succeeded, notably in Pernambuco on the north-east coast, confirmed the potential of Brazilian sugar. Like the earlier migrations of sugar, this new transatlantic version was made possible by the movement of men, skills, finance and crops from older sugar-growing regions. Brazilian sugar took root and thrived on the back of the manpower, experience and finance from Europe and from the Atlantic islands. When royal control was finally and securely settled in Bahia, in Salvador, the local sugar industry was able to thrive in relative safety. All this, however, was at great cost to local Indian peoples. To enable sugar to thrive, they were moved from the rich lands they occupied and regrouped into villages under Jesuit care.

At first, Brazilian settlers used Indian labour and tried to lure Indians from the interior to their coastal sugar settlements, but no amount of cajoling and persuasion could secure local labour in the numbers required. By the 1570s, Brazilian planters were turning, like the sugar planters on São Tomé before them, to African slaves for their labour force. By 1600, more than 200,000 enslaved Africans had been shipped to Brazil. Once the Portuguese colonised Angola in 1575, they began to develop their own slave trade from Luanda. No one at the time could have predicted the astonishing outcome. This early South Atlantic trade evolved into a stunning enforced migration of Africans. All told, 2.8 million Africans were to leave Luanda as slaves bound for the Americas, overwhelmingly to Brazil.[5]

The earlier settlements of some of the Atlantic islands had involved the enforced removal of people, by commercial and political interests, to cultivate sugar cane in distant lands. But Brazil set in train an altogether different process. There were, in effect, *two* processes at work in the late sixteenth century in Brazil, and both were to recur throughout the colonial and early national periods of history across the Americas. First, the indigenous people were removed to clear the land for settlement and cultivation; and second, foreign labourers were shipped in to bring that land into profitable cultivation. Indian peoples were driven from their lands, and Africans replaced them. In Brazil in the late sixteenth century, the numbers involved were small. But they formed, in essence, the initial landing party of a subsequent migration that was to transform the demographic face of the Americas. It was the beginning of the Africanisation of swathes of the continent. And this process had its origins – and its culmination – in the urge to cultivate sugar cane.

Despite all the other commercial possibilities, sugar quickly became Brazil's main export crop and it was not to be dislodged until the nineteenth century. Brazilian sugar began to arrive in volume in Europe in the mid-sixteenth century, initially via Lisbon and other Portuguese ports. By the end of that century, sugar was being shipped direct to northern Europe, especially to Antwerp (later to Amsterdam). The sugar economy of Antwerp and Amsterdam (founded on imports from São Tomé) now thrived on Brazilian sugar, which also landed in Hamburg and London.

The sugar previously grown in the Atlantic islands had passed through some simple stages of refining before being shipped to Europe, where sugar refineries took the raw sugar and refined it further into the lighter-coloured sugars most sought after by European consumers. Sugar refineries had been simple in medieval Egypt, and Marco Polo visiting China had noted them. In Europe, they were initially concentrated in Antwerp (before the city was sacked by Spanish forces in 1576). That city was the centre of the dynamic economy of the southern Netherlands (the centre of what was known as 'the rich trades') and served as a major distribution point for exotic goods imported by Portuguese shippers (though often financed by German, Jewish and Italian backers). Sugar was, then, just one of the many imports from distant markets, but sugar refining, with its unpleasant smoke and pollution, soon became a feature of Antwerp's urban landscape. In 1550, there were nineteen sugar refineries in the city.

Twenty-five years later, London, too, had become a major centre for refining sugar, along with other major European ports, as their merchants became increasingly involved in the sugar trade. By the mid-seventeenth century, Amsterdam boasted forty sugar refineries, despite local attempts to curb the coal-based pollution belching from the refineries' chimneys.[6] By then, the sugar was arriving in Europe from plantations in the Americas.

These were the years of ascendant Dutch trade and maritime power, and much (perhaps most) of Brazil's sugar was shipped across the Atlantic in Dutch ships. The sugar headed for northern Europe mainly because Brazil lacked the facilities to refine the sugar properly. In fact, we can gauge the exports of Brazilian sugar by the proliferation of refineries in Antwerp and Amsterdam. Amsterdam's sugar refineries increased from 40 to 110 between 1650 and 1770; London had 80 refineries by 1753.

In Brazil, the initial, crude system of refining created an important supply of rum which was consumed locally, although much was also shipped across the South Atlantic in return for yet more slaves.[7] And while rum also found an eager market in Europe and especially in colonial North America, it was sugar which was the engine of this remarkable slave-based system. Until the 1630s, Brazilian sugar had no real competitors. Thereafter, however, new British and French European settlements in the Caribbean islands created serious competition. Sugar from St Kitts, Barbados, Jamaica and, eventually – and most significantly – from the French colony of St Domingue, transformed the sugar economy of the Western world. But all of them, from Jamaica in the north to Salvador in the south, depended on imported African labour. The pattern was set: sugar meant slavery.

Although they came late to the imperial game, both the French and the British began to acquire tropical colonies in the knowledge that they contained great potential for commercial prosperity. Who knows what the luxuriant lands of the Caribbean might yield? And while Brazil had clinched the case for the development of sugar, there were other commercial possibilities awaiting settlers on the far side of the Atlantic.

The Spaniards brought a great variety of plants to the Caribbean, although not all their experiments succeeded (wheat was a notable failure), but sugar was to transform the region, despite it struggling at first against other commodities. The settlement of the smaller islands of the Eastern Caribbean was followed by attempts at varied forms of agriculture, notably tobacco, indigo and cotton. The early settlers, led by the Spaniards, at first 'dabbed and dabbled, trying one thing after another. Quite quickly, however, they turned to sugar growing and sugar-making.'[8] Sugar ushered in a fundamental and far-reaching human and ecological revolution. This was the 'sugar revolution' that was to transform the face of the Caribbean; it changed for ever the natural habitat, the face of the land and the very people who inhabited the region. Though we can calculate the changes in that arc of islands – the change in populations, the transformations in the local flora and fauna – we need to add to this complex formula the impact made on taste throughout the Western world (and later, throughout the world at large). Brazil had whetted the appetite of the Western world for sugar, but it was the shift to sugar in the Caribbean that inaugurated the corruption of the world by sugar.

Sugar was first cultivated on smallholdings, later on larger plantations. At first, the labour was mixed – free or indentured European labour, which increasingly became dominated by enslaved Africans as plantations took root and grew to huge proportions. Sugar cultivation, and the extraction of juice from sugar cane, were, by then, all well-known agricultural and industrial procedures, and Brazil had already shown that it was a commercial venture which, with luck and good management, could be a lucrative business. Settlers in the Caribbean adopted the systems established by pioneers in Brazil and used European financial backing and markets. At first, they remained uncertain which agricultural route to take (not unlike the settlements in North America), but the drift to sugar effectively began in Barbados in the 1640s, and spread quickly to Guadeloupe, Martinique, St Kitts, Nevis, Antigua and Montserrat, and Jamaica. Everywhere, the success of sugar depended on imperial protection – and good prices in European markets.

The eventual outcome was the emergence of the Caribbean as the world's centre for sugar production. Barbados exported 7,000 tons of sugar by 1650. Fifty years later, sugar production in the British islands had reached 25,000 tons, and now surpassed Brazil's 22,000 tons. In 1700, there were ten sugar exporters, all of them colonies in the Americas, dispatching 60,000 tons to the world's markets. All of them relied on enslaved Africans.

Within less than a century, the number of sugar exporters doubled, and they were producing 150,000 tons by 1750. In 1770, more than 200,000 tons were produced (90 per cent of it from the Caribbean). Two colonies alone – Jamaica with 36,000 tons, and St Domingue with 60,000 tons – accounted for half of the Caribbean sugar production.[9] None of this could have happened, on such a scale, without the unprecedented and brutal transportation of millions of enslaved Africans. Sugar had become synonymous with slavery.

* * *

Although sugar could be cultivated by small farmers, and their cane handed over to others for crushing and processing, it had quickly become clear that the best results, and the maximum returns on sugar, were on large plantations. But they were labour-intensive and required gangs of labourers working to a rhythm – and with an intensity – that was unlike other forms of agricultural work. The early sugar system of using mixed labour had effectively disappeared from the Caribbean by the mid-eighteenth century. European indentured labourers simply faded away and sugar was now dominated by plantations, and the plantations were populated by African slavery. There were simply not enough indentured labourers to satisfy the voracious labour demands of the sugar plantations.

Throughout the Caribbean, sugar planters came to prefer African slave labour. For a start, few challenged the morality of slavery itself (until that is, the upsurge of abolition sentiment – and the related law cases in Britain – in the last quarter of the eighteenth century). Moreover, types of slavery had existed in the Americas from the early days of European settlement, notably the small-scale (and largely unsuccessful) enslavement of Indian peoples. Africans shipped across the Atlantic – if they survived the crossing and landfall (and many did not) – belonged to a sugar planter for life. So, too, did their children. In the eyes of colonial and metropolitan law, African slaves were *things* – chattels, objects, property – to be bought, sold, inherited and bequeathed like any other item. This property status was the basis of black slavery throughout the vast colonial lands of the Americas and, despite the obvious and inevitable confusions (especially in the law), it defined Africans and their descendants throughout the era of slavery. But it also meant that slave owners could effectively do with them as they wished. Everyone involved in the increasingly complex Atlantic sugar trade – notably planters and traders – devised a multitude of reasons for using African slave labour, such as their natural strength for hard work, and their resistance to tropical ailments. But the simple truth was economic – Africans could be recruited at relatively low cost (despite the vast distances travelled), and they could be easily replaced. This was not true of other forms of labour.

So it was that slave ships, from all corners of the Atlantic system, from all of Europe's major ports, and from a number of American ports, from Rhode Island to Rio, converged on the Atlantic coast of Africa, where they exchanged a huge variety of goods for cargoes of enslaved Africans. They, in their turn, endured the uniquely terrifying Atlantic crossing, often having been incarcerated for many months on a slave ship in a port on the African coast as the manifest of slaves was filled. This maritime experience was brutal, disease-ridden, marked by horrific death rates and the capricious brutality on the part of the crew, who lived in daily fear of their African prisoners. Moreover, all this took place *before* the enfeebled survivors stumbled ashore to begin (if they survived at all) a lifetime's labour. Millions of them were destined for the sugar fields.

Until the 1840s, it was the African slave who, above all others, was the pre-eminent pioneer in the Americas. This Africanisation of the Americas was, to the modern eye, the most obvious transformation in Brazil, North America and in a string of islands that arced more than 2,000 miles from the southern tip of Florida to the north-eastern tip of South America. The numbers involved are simply staggering. We know that more than 12 million Africans were loaded on to the slave ships, and that more than 11 million survived to reach landfall in the Americas. What began as a transatlantic trickle grew into the largest pre-modern enforced movement of people the world has ever seen. In the 1570s, some 2,000 Africans were transported every year to the Americas. That had grown to 7,000 a year in the early 1600s, and 18,000 by the 1660s. The coming of sugar, however, saw those figures increase astronomically. By the 1790s, 80,000 Africans were arriving each year, with the very great majority landed in Brazil or the Caribbean. More than one million landed in Jamaica, and almost half a million in the small island of Barbados. Even the tiny island of Dominica received 127,900 Africans.[10]

In time, Africans, and their children born in the Americas, undertook every conceivable kind of labour, from the dockside loading ships' cargoes and processing the arrival of new Africans, through to the very edges of the American frontiers. They became miners, cowherds, lumberjacks and nurses, cooks and seamstresses, and many were skilled craftsmen – joiners, wheelwrights, metalworkers, boilermen in the sugar factories, drovers in the fields and domestic servants in towns and in plantation houses. Enslaved Africans were inescapable throughout the Americas. Above all, they were to be found in the largest concentrations labouring at the back-breaking work in the sugar fields.

There they were marshalled into 'gangs' – outsiders used military imagery to describe slaves working in sugar – and the first gang of the fittest slaves (both men and women) were assigned the heaviest work, cutting and hacking the cane. They were followed by the second and third gangs – men, women and children – who collected and bundled the cane and loaded it on to carts heading to the factory. There, skilled slaves took over the process of crushing, boiling and filtering the cane into the juices that ultimately became molasses and barrels of crude muscovado sugar. All this was then transported by ox and mule carts to the waterside, and on to the ships destined for distant ports and refineries – and from there to the eager consumers of Europe and the wider world.

Slaves on sugar plantations entered the labour force as children, as soon as suitable work could be found for them. As they aged, they slipped from physical, demanding labour to other menial, less demanding chores. They worked from childhood to old age, or until incapacity, accident or sickness made them (in the unforgiving phrase of plantation ledgers) 'old and useless'.

Africans had become an investment long before they started work in the sugar fields. They had a value when first traded in Africa and had a price on their heads from the moment they stepped on to a slave ship. Thereafter, they were an item of trade, although there was an ironic element to this status. Because the Africans represented an investment (by slave traders, and later by planters), they needed to be managed and looked after. For all the cruelties, ill-treatment and punishments on the slave ships and plantations, slaves represented a costly investment by their owners. On the plantations, owners monitored their labourers' health and well-being, allocating them to appropriate tasks from infancy to old age, according to their physical age and condition. By the late eighteenth century, Jamaican sugar slaves were being periodically checked by a doctor – and even vaccinated against smallpox, which was a major scourge among the slaves. Planters also provided appropriate food, shelter and clothing, although all this was supplemented by the slaves' own efforts, working on their plots and gardens in their spare time. Such necessities were not provided for philanthropic or humanitarian motives; they were dictated by simple economic realities. To get the best from their enslaved labour force, sugar planters needed their assets to be as fit and as healthy as circumstances and finance allowed.

So it was that the human patterns emerged which are familiar to any student of sugar slavery, or to anyone glancing at contemporary illustrations. Images became commonplace of strong, youthful male and female adults toiling at the heaviest labours in the fields, with younger, less mature or older slaves following behind, and with a small band of skilled slaves transporting and processing the cane into sugar and molasses destined for the ships. On all plantations there were scatterings of old people and women with gaggles of children, undertaking all the domestic and local tasks in and around the houses, the yards and the animal pens. Whatever their tasks, slaves worked from dawn to dusk.

For their part, sugar planters had clear ideas of what they expected slaves to achieve, whatever their task. Slave owners and their managerial overseers developed a keen sense of what their labour force could accomplish, in a precise time frame, at a specific job. Their paperwork, compiled year after year, season after season, in the huge ledgers imported from Europe, provide an actuarial guide to what had been successfully achieved. The accountancy of the sugar trade was exacting and unrelenting.

The plantation ledgers provide a blueprint for running a successful sugar plantation. There had never before been such a calculated and remorseless analysis of land and labour, and never before such an unyielding system able to extract the maximum returns from the labour force. Through all this, slaves worked not simply to the tight discipline of the sugar season – cutting and processing the cane from January to mid-summer, planting new cane and tending the fields thereafter for the next crop – but they worked under the threat of severe compulsion. It was not an artistic fancy that when outsiders sketched or painted slaves at work in the sugar fields, they invariably portrayed master and drivers, in the saddle, whip in hand, ready to goad the labour force to work harder. True, material rewards were built into the system – extra food, clothing, treats on high days and holidays – but sugar was produced by an unforgiving system of brutal enforcement. Yet who ever gave this a moment's thought, or heard the sound of the lash, when spooning sugar into their tea or coffee in London or Paris?

The sugar plantations were highly organised systems. The enslaved labour force, like every acre of land, was tabulated and regulated. Plantation paperwork – the large ledgers that were the stock-in-trade of plantation bookkeepers – documented every aspect of plantation life. The lists of slaves – their names, ages, health and value – were all tabulated alongside the same details for the plantations' livestock and material objects. Everything had a cost and a value, from the tools in the carpenter's box to the ranks of Africans slashing their way through the swaying sugar cane.

Where planters lived abroad (it was the ambition of the more successful ones to retire, and go 'home' to France or Britain), the attorneys left in charge of a property dispatched bundles of detailed letters back and forth across the Atlantic explaining each and every move of the property and its inhabitants. This was not only a world of enslaved people, but it was an international world shaped and fashioned by literate people. Traders, merchants, shippers, ships' captains, planters and their scribes – all wrote copiously to each other, advising, ordering, buying, selling, instructing. The wheels of this Atlantic sugar trade were lubricated by a vast literate and numerate culture. In the British case, this was greatly helped by the remarkable role played by educated Scots, both in the sugar islands and in their metropolitan homes.

While it may seem perverse to modern eyes, few questioned this system of slavery. Not until the mid- and late eighteenth century did substantial, then widespread, doubts emerge about the morality of slavery and slave trading. The reason was straightforward enough: here was a system, driven forward by sugar, which yielded abundant wealth and well-being for everyone – except, of course, the Africans. Although the focal points of slavery were on the African coast, on the Atlantic slave ships and the plantations – and although Europeans might easily imagine that slavery was a distant, 'colonial' issue – its benefits and consequences were clear enough in the European heartlands. It was obvious on European docksides where ships were loaded and emptied, which ones were bound for Africa or landed with slave-grown produce. It was obvious in the array of industries that constructed the thousands of ships and filled them with a variety of cargoes bound for the African slave markets, and it was clear enough in the factories and warehouses which processed slave-grown produce, such as the sugar distilleries of Amsterdam and Liverpool, and the tobacco warehouses of Glasgow. And it was obvious in the black people who found their way to live and settle in Europe.

The benefits of slavery were at their most visible and impressive in the grand homes of successful slave traders and merchants (in Bordeaux, Bristol and Bath, for example) or in the rural mansions of the major sugar barons. And it was present, of course, in the simple pleasures of a sweet cup of tea or coffee in the humblest of homes. The fruits of slave labour had thoroughly permeated the Western world, and had become so entangled in the social and physical fabric of Western life that it was hard even to notice it. Who could doubt or question slavery when it brought such benefits and simple pleasures to the 'civilised' world? Slavery went unchallenged and unquestioned in very large part because it yielded such benefits and pleasures to so many people. Any pain or misery it inflicted on millions of Africans were largely invisible to Europeans, because they lurked somewhere over the horizon, out of sight, and out of mind.

Slavery thrived. Ever more Africans were loaded on to the Atlantic slave ships, ever more of them endured the hellish oceanic ordeal, and vast numbers of them were marshalled into the ranks of plantation labourers to endure a life of bondage – all for the benefit of a Western world about which they knew little, if anything. All went largely unchallenged effectively until the 1770s. Then a series of legal cases in England and Scotland, which wrestled with the problem of slavery in Britain itself, began to tug at the fabric of Atlantic slavery. Was slavery legal in England or Scotland? Was it legal to oblige a person to return from Britain to the slave colonies against their will? And, even more spectacular in its murderous consequences, could slave traders claim insurance for Africans thrown overboard from slave ships in times of shipboard distress?

These small points of law exposed both the reality of slavery and slave trading, both in the courtroom and among the wider reading and politicised public. All this, at a time of heightened radical agitation, ushered in by the American and then the French Revolutions, generated a new and unprecedented political and moral debate about slavery. The irony, however, was that this debate came at a time when the slave trade and slavery boomed. Before then, however, the benefits of slavery throughout the Western world had far outweighed any doubts about basic human suffering and oppression – criticism went unheard.

It was, for years, possible to minimise the negative aspects of slavery. The excesses of exploitation and oppression ushered in by modern industrial change in the nineteenth century seemed to overshadow what had gone before. Yet we need to remember that until the late eighteenth century, the sugar distribution network – above all of the Caribbean and Brazil – embraced some of the largest commercial enterprises operating anywhere in the world. They were the most capitalised, the most productive, and involved the largest labour forces.[11] It was no wonder, then, that all the European imperial powers valued their Caribbean sugar islands, and fought for them throughout the era of slave-grown sugar. The Africans had made possible the sugar revolution, and that revolution had made the islands the jewel in the imperial crown.

What this upsurge in Caribbean sugar production made possible was the massive consumption of sugar throughout the Western world. As Europeans settled distant colonial outposts and colonies – in North America, India, Africa and, after 1787, Australia, they took with them an attachment to sweetness. The end result was that sugar, produced by Africans in the Americas, then shipped to Europe for refining, was sold on wherever Europeans set up home, established trading or military posts – or simply camped on the very edge of the imperial frontiers. Sugar, once the privileged luxury of a wealthy elite, was now being consumed all over the world. By the end of the eighteenth century, sugar was available everywhere; it was sold in marketplaces, corner shops, fashionable emporiums and even in the simplest of village stores. The world had become addicted to sugar.

4

Environmental Impact

THE MASSIVE INCREASE in sugar output in the course of the eighteenth century was made possible *not* by improvements in agricultural or sugar processing, but by sugar planters moving on to new sugar lands. There were, it is true, some minor improvements in methods, but the real change was the simple expansion of sugar cultivation.

This expansion had a massive impact on the Caribbean environment. As new lands were brought into cultivation, the natural habitat was cut back and destroyed, usually by the favoured local system of 'slash and burn'. Here was the latest twist in the ecological transformation wrought by sugar – the destruction of great swathes of the natural environment, notably the local rainforest, and the creation of man-made orderly fields which formed the regular, geometrical shapes of plantation agriculture.

The 'sugar revolution' looks peaceful enough when we glance at the methodically surveyed and neatly arranged field systems of sugar plantations at their height, with their well-manicured fields and crops, paths and roads slicing through the countryside to make easier the movement of people and goods to and from the coast, and then to and from Africa or Europe. What is easily overlooked – for the simple reason that it had been *erased* – was the natural world that existed *before* the sugar revolution. The indigenous rainforests effectively disappeared, replaced by fields of sugar cane – orderly and growing or receding as the season progressed – which dominated a landscape which had, to the first arrivals, seemed dense and impenetrable. Sugar created a new natural world that seemed to have been brought into being by geometry: land apportioned into squares and rectangles, and all sectioned and carved up by walls and fences. It was a landscape created by humans, and shaped by generations of meticulous surveyors, their mathematical and technical skills reducing what had once been a teeming and seemingly impassable forest and bush into an orderly and manageable agricultural system.

Looking at a surviving sugar plantation today, the landscape seems natural. But, in 1750 say, it was new and revolutionary, an orderly vision of nature brought forth by the demands of the need to cultivate ever more sugar cane. In its wake, it left irreparable change and human damage, which was recognized even by the mid-1700s, when mahogany trees, themselves valuable for furniture-making in Europe, had been destroyed by earlier slash-and-burn systems to clear the way for sugar.

The change in human diet brought about by the rise of the sugar economy since *c.*1600 is easily described. But less well known are the dramatic human and environmental upheavals brought about by sugar. The human and physical face of the sugar-growing regions was transformed by the arrival of alien labour imported in huge numbers to work in the sugar fields. Sugar plantations – which quickly established themselves as the critical means of cultivating sugar – were also responsible for transforming the natural physical landscape. The appearance of the natural environment of sugar production seems unremarkable, and seems at first glance to be merely a reflection of the overall natural setting. So, too, the local populations. In fact, both the human and physical face of sugar regions had been brought into being *specifically* to cultivate sugar. Sugar brought about an upheaval in both the environment *and* in the nature of the people working within that environment.

The post-Columbus settlement in the Americas by alien humans, animals, flora and fauna set in train a complex process of human and natural upheaval. Best known, and the most obvious, were the disasters which befell the native peoples. That started long before the onset of the 'sugar revolution', but the coming of sugar, especially the rapid development of large sugar plantations, rounded off a process of seismic human and environmental change heralded by the first European landfall in the Caribbean in 1492. A century later, Las Casas wrote perhaps the most poignant contemporary account of what happened to the people of the region:

We should remember that we found the island full of people, whom we erased from the face of the earth, filling it with dogs and beasts.[1]

The Taino people, who had settled the Caribbean islands from South America, had brought about no great upheaval to the local ecologies. The Europeans, however, brought something totally different. Theirs was an invasion which viewed native

peoples as a major obstacle to successful settlement and colon-
isation, and the animals, plants and systems they introduced
utterly transformed the island's ecological systems.

The exact numbers of people in the Caribbean on the eve of
the European arrivals remain elusive, but the broad outline of
what happened is undisputable. On the eve of the European
invasions, all of the larger Caribbean islands were populated
and productive. Hispaniola may have contained more than
1 million people, Puerto Rico upwards of half a million. Cuba
between 100,000 and 150,000, Jamaica fewer than 100,000,
with 40,000 scattered around the Bahamas. There were some
2 million people living in the Caribbean.

Within less than a century, all had vanished. This population
was comfortably adapted to the resources and the environment
of the islands, and had readily adopted plants and knowledge
from South America. True, they had faced periodic natural
disasters – volcanoes, earthquakes, even tsunamis (1498 and
1530) – although hurricanes, as they are today, were the most
common. Yet all these natural dangers were as nothing
compared to what followed the arrival of Columbus in 1492.
What seemed 'a puny human intervention . . . was to prove
catastrophic'. The finest modern historian of the Caribbean
describes the process under the blunt heading 'Columbian
Cataclysm'.[2]

The years after 1492 saw the arrival in the Caribbean of
peoples from Europe and Africa, along with their plants,
animals and technologies – some of which originated even
further afield. Soon, Africans were being fed in the Caribbean
islands on fish from the North Atlantic to enable them to cultivate crops from North America, Arabia and Asia – and all to
flatter the taste of the Western world. New animals, and new

agricultural systems which demanded large expanses of land – much of which was originally rainforest – transformed everything. Crowning all this – literally, in the case of European monarchical government – the political control imposed on the Caribbean peoples was a form of violent, military power totally unknown by the indigenous populations. Again, Las Casas captures the point:

> *How can a people who go about naked, have no weapons other than a bow and arrow and a kind of wooden lance, and no fortification besides straw huts, attack or defend themselves against a people armed with steel weapons and firearms, horses and lances, who in two hours could pierce thousands and rip open as many bellies as they wished.*[3]

The Caribbean islands became in effect the crucible for the creation of totally new cultures and peoples. The Taino peoples disappeared, their place taken by Creole cultures dominated by Africans but controlled by Europeans. And they were driven forward, from the 1640s, by the engine that was the sugar industry.

The pioneering settlers in the islands had first to scratch a living from the land they brought into cultivation. But the cultivation of export crops – tobacco and cotton at first, then sugar – demanded a herculean task of clearing scrub and forest. Axes – and axemen – were in short supply and the easiest and most common system was simple slash and burn. Both in St Kitts and Barbados, it was slow progress for twenty years. The arrival of sugar and the creation of the first small plantations and landholdings hastened the process. As sugar thrived, and as export increased (in parallel with the numbers of imported

Africans), ever more land fell victim to the march of sugar cultivation. With sugar cane, and experience from Brazil, sugar took hold in Barbados by 1640. The first sugar planters made good profits from their sugar, invested in more Africans, more land and more sugar cultivation. Good, useful timbers were kept for construction and export, but large tracts of woodland were simply destroyed to create cultivable land.

By 1650, much of the forest in central Barbados had been destroyed. A mere fifteen years later, all but the most inaccessible gulley and hillside on the island had been cleared of their forest. For the first time, the landscape of Barbados was open. Visitors sailing along the coastline, or riding into the interior, could see mile after mile of sugar plantations. In the words of Richard Ligon, writing in around 1647: '*As we passed along near the shoare, the plantations appear'd to us one above another.*'

The ecological problem was felt even by the planters. Having cut down all the trees, they had to import coal from England to boil and process their sugar crops.[4] As sugar thrived, the more successful planters were able to buy out their smaller neighbours. Bigger sugar properties began to dominate the physical landscape just as the bigger planters began to dominate the social and political landscape. By the end of the seventeenth century, sugar was king, and the king's voice was heard – and respected – in London. Yet none of this would have been possible without the Africans; by 1700, some 180,000 had been landed in Barbados.[5] For all the profit and well-being this brought to the planter (and the imperial treasury), the loss of the forest had a deeply damaging impact on the island itself. Alien trees were planted and thrived, of course (coconut, guava and a host of shrubs); but so, too, did rats, who feasted on the sugar cane and became a major plague throughout the sugar-producing islands.[6]

What laid waste to swathes of the islands' natural vegetation was fire. The French burned out the English in St Kitts, as well as the local vegetation in 1666–67. They did much the same in St Croix. By 1672, the forests on the lowlands of St Kitts, Nevis and Montserrat had gone up in smoke. Antigua alone seemed to have untouched areas of virgin forest. But even there, fire had consumed tracts of the island's cover. It was no surprise that a similar pattern of land settlement and deforestation took place on other islands when local settlers turned to sugar cultivation. Jamaica, for example, was transformed in the eighteenth century. Similarly, the sugar boom of the eighteenth century saw deforestation in Guadeloupe and St Domingue. In Antigua, sugar cane dominated every district by 1750, and very little forest remained. As Jamaica overtook Barbados sugar production by 1712, its forested landscape, too, had succumbed to plantation vistas.

Contemporary maps reveal how the early settlement along the coastal plains and inland valleys was augmented by settlement on interior locations – wherever sugar cultivation could be wrested from the luxuriant cover that formed Jamaica' natural wilderness. The number of Jamaican sugar mills, for instance, increased from 57 in 1670, to 419 in 1739, to 1,061 in 1786. In eight years in the 1790s, 84 new sugar estates were established in the northern parish of St James alone. Not only that, but, as in Barbados earlier, sugar estates expanded significantly. In 1670, 724 sugar planters worked an average of 261 acres. By 1754, the average acreage had grown to 1,045 acres, and 4 per cent of planters owned land in excess of 5,000 acres.[7] All this land had been converted to sugar cultivation from native forest by slash and burn, and by the back-breaking toil of African slaves. Some 95,000 had landed in Jamaica by 1700. In the next century, more than 800,000 would be landed, although large numbers would eventually be shipped on to other colonies.[8]

The pattern was similar wherever sugar told hold. The traditional wilderness and forest was replaced by neatly ordered estates; cane fields were sliced and arranged around paths and roadways leading to the local factory, and from there out towards loading points at the nearest waterside for onward shipment to Europe and North America. And the whole was managed by small cadres of white owners and their literate and numerate staff, and their hordes of African slaves. We are told that planters:

> . . . Buy them out of the Ship, where they find them stark naked, and therefore cannot be deceived in any outward infirmity. They choose them as they do Horses in a Market; the strongest, youth-fullest, and most beautiful, yield the greatest prices.[9]

From the Chesapeake Bay to Charleston, throughout the Caribbean and onwards south to Rio, a steady stream of ships, from all the major European maritime nations and from the Americas, disgorged their human cargoes of enslaved labourers to push back frontiers and bring the land into cultivation. Though slaves worked in every conceivable occupation – from sailors to cowboys – the great majority were destined, at some point in their lives, to be plantation labourers.

In time, plantations developed a huge variety of agricultural activity: tobacco, rice, coffee, cotton. But it was with sugar, first in sixteenth-century Brazil, then in the Caribbean in the seventeenth century, that the plantation came to its most complete, profitable and ideal form. But it did so at enormous cost to the native ecologies of the region, and to its enslaved African labour force, though this mattered little to the men who owned, managed, supported and profited from it.

The sugar plantation became a highly successful blueprint of how to turn virgin land and an enslaved labour force to everyone's financial advantage – everyone, of course, except the slaves themselves. It was as if the settlers had, in the sugar plantation, devised a cornucopia which spilled forth prosperity – King Sugar – on an unparalleled scale.

The ending of slavery, however, created a serious problem for sugar planters. Freed slaves looked on the plantations as houses of bondage and did not want to work on them, and simply quit. The labour vacuum was filled by a new system of 'indentured labour' recruited in India and shipped, through Calcutta, to the former slave colonies. Indian labourers – not quite slaves, but certainly less than free – arrived in huge numbers. The African diaspora had been replaced by an Indian diaspora.[10] The mass movement of peoples in the years after slavery were from the Indian subcontinent. And most of them worked on plantations recently vacated by freed Africans, while others laboured in newer colonial settlements, where free labour was unavailable or unwilling to embark on the lifetime of drudgery that plantation labour entailed.

What had been pioneered and perfected in the sugar fields of the Americas was later transferred to many other regions, in very different agricultural activity. The cotton boom which transformed the USA in the first half of the nineteenth century was anchored in the slave plantations of the South. When the British introduced tea into India, it, too, found a home on plantations.

Much the same happened in East Africa with tea and coffee, in West Africa with palm oil and cocoa, in Malaya with rubber, in Hawaii with sugar and pineapples, and in Fiji with sugar. In all these cases, and more besides, two related elements brought about massive change. The natural, existing landscape was overthrown by the introduction of a dominant alien crop for cultivation. Land was cleared, forests and other local habitats destroyed, and native peoples were subjugated to the rigours and alien dictates of plantation labour. And when local labour could not, or would not, bend to the dictates of plantation life, plantation owners and governments that backed the entire enterprise sought labour elsewhere. Once again, labour was moved enormous distances to feed the voracious appetite of the plantations. Indians were dispatched to the Caribbean islands, to the sugar islands of the Indian Ocean, to Ceylon's coffee and tea plantations, and even further afield to the cane fields of Fiji.

Wherever sugar was planted, it set the mark of Cain on the landscape. Sugar could be grown differently – sugar regions are, after all, characterised by untold numbers of small 'cane-farmers' working on just a few acres – but what happened in the Caribbean from the seventeenth century onwards established the dominance of the plantation. Those plantations demanded labour on a mass scale and workers were subject to the most arduous of conditions. The sugar plantations worked by African slaves were a realm unto themselves, yet were linked vitally to distant worlds. It was also a malleable and resilient organisation which flourished even after slavery had gone.[11] It thrived in all corners of the tropical world from the days of slavery down to the present day.

What today seems natural – the orderly Caribbean landscape, home to people of African, Indian and European descent (and mixes of all of them) – is the product of particular historical circumstances. And the engine behind that historical process, an ecological and human transformation of enormous proportions, was the sugar plantation and the growing world appetite for sweetness in its food and drink.

From one Caribbean island to another, a twin process transformed the natural and human face of each location. As the population became increasingly African, the original, natural environment succumbed to the march of sugar cultivation. And this happened even in places where sugar was not the dominant local crop. In fact, the two processes were linked, because it was the vast numbers of imported Africans who were used as the shock troops not merely for sugar cultivation, but also for the transformation of the land itself. It was Africans who hacked away at the bush, who burned the rainforest and took axes to the blackened stumps, and it was they who then tilled the fields and brought them into orderly shape, creating space for the cultivation of sugar and other export crops. And for all those Western consumers of sugar, the slaves' plight – and that of the land they were compelled to change for ever – was very much a world away. Out of sight, and very much out of mind.

5

Shopping for Sugar

WHEN THE MODERN British Co-operative Movement was founded in 1844 in Rochdale, the founding members, a group of local weavers, set aside £28 to buy the first essential groceries for their enterprise. Alongside the butter, flour and oats, they bought some sugar. By then, sugar had become a basic of everyday life, available in various forms, cheap and costly, across the world. But how did people acquire their sugar, a commodity grown and produced in distant corners of the world? How did they shop for sugar? Today, in a world which often seems shaped and paced around a culture of shops and shopping, it is hard to envisage how earlier societies acquired life's essentials – and its luxuries.

Before the emergence of modern shops, people of all social strata depended largely on local goods. They traded, cultivated or fabricated life's essentials in the locality or region. Shops as we know them are relatively new phenomena, so where and how did people acquire sugar – and other commodities – which were not local but which had been brought thousands of miles to the consumer?

Those who were prosperous had labourers and artisans to cultivate and prepare their food, and to make their shoes and clothing and their household essentials. Poorer sorts scratched a living as best they could, clothing and feeding themselves by their own labours. But everyone needed a local trading place – a market – or itinerant traders for foodstuffs and clothing, or to acquire materials to make into clothes and shoes. Most of this activity was very local, though a range of items travelled relatively long distances to sustain family and community life – timber and coal, salt fish from the nearest coast and, for the rich, luxury items from further afield, such as wines, oils, spices and costly textiles from France and Italy. All these goods found their way to the local market, and the marketplaces formed a spider's web of distribution across the face of medieval Europe.

Markets were at the heart of urban and civic life. They were regulated and inspected, and they occupied a vital role in local and regional social and economic life. Thousands of them linked towns to their rural and agricultural hinterlands, and even to the wider world. In the years 1200–1349, an estimated 2,000 new markets were established in England alone. The medieval market square became – and in many places has remained to this day – 'a defining feature of the urban landscape'.[1]

On market day, the place buzzed with commercial and social life. Butchers and fishwives offered their wares alongside linen drapers and tanners. At first, market traders used temporary stalls which were then removed when the market closed but, by the mid-thirteenth century, many of these had become permanent structures. In time, market stalls became more elaborate and, led by fishmongers and butchers, evolved into small, permanent buildings – shops, or 'shambles' – where foodstuffs would be protected and prevented from spoiling. (York's famous Shambles survives to this day, though is now regarded more as a tourist attraction, and is perhaps the best example in England.) Gradually, open-air trading places had begun to give way to enclosed trading halls, and to enclosed shops.[2]

Most customers went to markets to buy their goods, but some items – small-scale, sometimes affordable luxuries – could be bought from travelling peddlers. Rival traders tended to dislike peddlers, who had acquired a bad reputation, despite fulfilling an important role; their travelling boxes and satchels contained a range of cheap goods which people wanted. By the late sixteenth century, peddlers were the conduit for providing customers with more exotic, luxury goods from distant places, tobacco being an early example.

At the more lavish end of the scale were the fairs which had flourished in the twelfth and thirteenth centuries. By the end of the Middle Ages, there were an estimated 2,700 fairs, some of them very specialised (such as goose, cheese or horse fairs – again, some of which survive today), but others brought together commodities from across the country, and from distant places in Europe. People travelled great distances both to sell and to buy at the fairs. European merchants came to buy, bringing with them goods which had travelled north from the markets of southern Europe and the Mediterranean, crossing the Channel with olive oil, furs, wines – and now sugar – from distant European markets.[3] It was at such fairs that the stewards from royal, aristocratic and religious houses stocked up with luxury goods and foodstuffs. Here, too, the more impoverished shoppers first encountered luxuries they might covet but could only dream of buying.[4]

Fairs were important for larger households that needed to store up essentials and luxuries to cater for the large numbers living there, or for the large groups who might be invited to visit and dine. Fashionable travellers to London, or their servants travelling on their behalf to London's major fairs, bought and placed orders for exotic goods that were unobtainable in rural and provincial society.

By the late Middle Ages, however, fairs went into decline, although many of the exotic items they once offered could now be more easily obtained in London. Sugar, as we have seen, had found its way into the major households across the country, and whatever route it took – via fairs, markets or London outlets – substantial supplies of sugar were stored in aristocratic households by the late thirteenth century. Two such examples were the kitchens of the Countess of Leicester in 1265 and the Bishop of Swinfield in 1289, both well stocked with sugar.[5]

By then, major English cities boasted an array of shops. There were, for example, 270 in Chester by 1300. Cheapside in London alone had 400, and even small towns had their own shops. They often clustered close to the town's marketplace, while others grew up close to river crossings, or along major thoroughfares, or at important street crossings wherever people had to congregate and gather. What had emerged, using today's parlance, were shopping centres, although the shops looked nothing like their modern counterparts. In the main, they were little more than the front room in a house, sometimes with a low window sill which opened and acted as a counter for the customers outside. Shop windows came much later, with the development of new, cheap glass, and a new style both for the promoting and selling of goods. Most shops, especially outside London, remained small and cramped, and although they might specialise in a certain line of goods, they often sold a fair range of items.

Sugar began to appear at these outlets in the late Middle Ages; it was available in London grocers' shops in the late fourteenth century. But it was the development of the sugar plantations, first in the Atlantic islands, then, more spectacularly, in Brazil, which saw sugar finding its way in substantial volumes on to the shelves of English shops. As the amount of slave-grown sugar increased, the price of sugar fell across Europe, and sugar became a common item in shops up and down the land. It was, of course, readily available in the major port cities of London, Bristol and Liverpool, but quickly spread out across the country.

London – with its trade links to European ports and markets and, by the late sixteenth century, onwards to a wider world – became *the* marketplace for all sorts of exotic produce. It did not, however, yet rival the scale of Lisbon and Amsterdam, whose long-distance trading vessels brought the exotic commodities of Asia, Africa and the Americas to local consumers. Many of those goods were traded onwards from Lisbon and Amsterdam to merchants who shipped them into England – especially to the south and east of the country. Sugar was among the earliest and most valued of those commodities and, as more and more of it flowed from the slave plantations, sugar began to appear in the humblest of shops across England.

Among the stock left by a London grocer in 1573 were 'seventeen sugar and five candy chests'.[6] By the early seventeenth century, sugar was even available in the smallest and remotest of towns; it was sold in Macclesfield in 1635 and

Rochdale in 1649. An ironmonger in the Cheshire village of Tarpoley in 1683 also sold sugar and molasses.[7] Fifty years later, the inventory of goods owned by Richard Johnson, a Kentish 'tallow chandler' contained a 'small parcel of penny sugars'.[8] Sugar was now available everywhere, and was sold in what today seems unlikely outlets.

By the late seventeenth century, sugar had become a ubiquitous feature of the social and dietary life across Europe. Grocers who traded with wealthy English customers in the seventeenth century stocked the luxuries they craved, notably coffee and, increasingly, tea – but above all they sold sugar. Some merchants went to elaborate efforts to ensure that sugar reached deep into the most remote rural heartlands of their trade. When Thomas Wootton, a grocer in Bewdley, Worcestershire, died in 1667, his trade in sugar (and other commodities) extended into six counties. Distant shopkeepers owed him money for goods he had dispatched to them. Much further north, Braham Dent of Kirkby Stephen in Westmoreland secured his supplies of sugar from across the north of England, and even as far south as London.

These two provincial merchants were just small, regional examples of a nationwide pattern of trading links. A grocer or merchant, often close to a port or major city, distributed goods far into the interior and hinterland.[9] Shopkeepers, merchants and traders bought, sold and borrowed, sending packages via trusted friends and colleagues. Quakers fared especially well here, because they were known to be reliable; shopkeepers and merchants trusted them with goods and with credit – knowing that their word was their bond – and that they would always pay. Hence the success of the major Quaker businesses from the late eighteenth century onwards.[10]

By the end of the eighteenth century, shops were to be found even in small, rural communities. 'Country shops' were now selling what had, a century before, been 'luxury goods' and even the rural poor had come to expect their sweet drinks. By then, too, in the bigger towns, more successful grocers' shops had evolved into a recognizably modern form, with lavish interiors, counters, drawers, canisters, jars and costly wooden fittings to house sugar and other goods.[11] And in all likelihood, the rich mahogany often used in such fittings was yet another by-product of the sugar industry, having been cut down and dragged to the ships by teams of African slaves. Shop signs and trade cards announced a shop's sugar business via images of sugar loaves.[12]

By the mid-eighteenth century, sugar was everywhere. It was part of a family's regular shopping routines, and a major ingredient in their everyday diet. Fashionable shoppers in York could buy their sugar (in 1766) from Nicholas Sequin, a French confectioner, who sold a large range of luxuries – 'Confits of all Kinds' – alongside sugars, pastes and cakes, various syrups, and sweet decorations to grace fashionable dining tables.[13]

For poor customers, a local grocer chipped a few ounces from a crude, cheap sugar loaf, grinding it into granules before wrapping it in paper. For the more discerning and prosperous, sugars were branded by the name of their island or place of origin. Customers had obviously learned to appreciate distinctions between sugars, and to know the better sugars from the poorer kinds. Sugar could be bought as a powder, in lumps, in loaves, raw or moist, or it was known by its place of origin – Barbados, Jamaica or Lisbon (for Brazilian).[14] As the eighteenth century advanced, and as the slave islands disgorged ever more sugar destined for Europe, sugar (alongside coffee, then tea) came to form a substantial part of what people spent on their weekly groceries. When women shopped and bought goods from a grocer, they were likely to buy sugar – along with tea.[15] Anyone unsure *where* to buy sugar could always head to the nearest apothecary shop. There, it was displayed in jars, bottles and chests with the name 'SUGAR' emblazoned on the front. In time, such bottles and jars were even displayed in the window to catch the attention of passers-by. Much the same was true across Europe. In Geneva, a local apothecary shop displayed sugar in an attractive porcelain pot – '*sucre candi*'.[16] Here was an ingredient to be mixed with other medicines and cures, to sweeten a bitter taste or, according to contemporary medical wisdom, to impart physical qualities of its own. Sugar had established itself as a basic ingredient on the shelves of the apothecary shop everywhere by the late eighteenth century – a reminder of sugar's ancient role in medicine – but, more importantly perhaps, it was now universally available as an ingredient in the food and drink of people in all sectors of society. A commodity produced 5,000 miles away by Africans labouring on plantations in the Americas was now a staple on the shelves of the most commonplace of shops in the remotest of communities – an indispensable aspect of everyone's daily diet, and an ingredient dispensed by England's apothecaries.

Not only was sugar now available in abundance for rich and poor via all sorts of shops and outlets, but there was another, related feature of the West's addiction to sugar which tends to go unnoticed. So common had sugar become in domestic life that, by the mid-eighteenth century, manufacturers were producing sugar bowls by the tens of thousands. The sugar bowl became a common object on dining, coffee and tea tables on both sides of the Atlantic. The Chinese porcelain industry

(Europeans did not know how to work with porcelain until the 1720s) disgorged hundreds of thousands of items for the West's appetite for sweet tea and coffee. In the process, they also learned to provide sugar bowls to match. European (later North American) craftsmen and manufacturers copied the Chinese in whatever material suited the market – porcelain (after Europeans learned how to make it) and silver items for the rich, pewter and basic pottery for less affluent customers. Fashionable eighteenth-century houses now displayed costly and incredibly beautiful porcelain sets from Sèvres, Meissen and Dresden, and later from Worcester, Derby and Wedgwood. And all of them manufactured sugar bowls. Today, the most beautiful are displayed in the world's museums and palaces, delightful reminders of the way sugar had established itself in Western life.[17]

Poorer members of society made do by wrapping their sugar in paper provided by a shopkeeper, although even they were sometimes able to acquire their own sugar bowls, the chipped, cracked and damaged versions handed down by their superiors, normally via servants. Eventually, even the previously exalted status symbol – the sugar bowl – like second-hand clothes and footwear, found its way into the poorest of homes.

Around the fashionable tea and coffee services there evolved major social routines of drinking sweetened tea and coffee. Such rituals were at their most striking in spa towns (most notably Bath), and at those watering holes close to Europe's major cities, and they became a retreat and an escape from the dirt and noise of hot summer days in a city. The bric-a-brac for all this – tea and coffee sets – could be bought by the end of the eighteenth century at shops specialising in tableware. In London, Josiah Wedgwood's shop was perhaps the most famous, where its location and design, allied to his new marketing techniques in Britain and abroad, elevated his products to become a global phenomenon. Alongside his teapots, cups, saucers and plates, Wedgwood's sugar bowls graced the tables of the prosperous from Russia to Portugal, from North America to the Caribbean. His cheaper wares, specifically aimed at the middle strata of society, ensured that even modest homes boasted the appropriate tableware. And none was complete without the ubiquitous sugar bowl.

Today, we don't even notice sugar bowls filled with sugar; they are a standard and unremarkable feature in most households, cafés and restaurants. The simple (or costly) sugar bowl had emerged as an item of everyday life, courtesy of the development of shops and shopping.

Again, though, those who coveted and purchased their decorative sugar bowls did so oblivious to the plight of those enslaved Africans who were fuelling the industry, and who had originally cultivated the sugar in distant tropical lands. Sugar had established its importance for the way it sweetened a variety of foodstuffs, making even the blandest of dishes more appealing, but, above all, it had become the agent which made hot, bitter beverages palatable to Western taste. What, after all, could be more British than a sweet cup of tea?

6

A Perfect Match for Tea and Coffee

T HE RATE OF coffee consumption in the modern world is truly astonishing. One estimate in 1991 was that seventy-six cups were drunk for every man, woman and child on the planet.[1]

Yet coffee's popularity was established on a global scale only three centuries ago. Moreover, it owes much of that popularity to sugar. Or perhaps we should think of it the other way round?

Sugar became popular as the natural partner of hot beverages – coffee and tea (and, to a lesser extent, chocolate). All are consumed in their native regions as bitter drinks – China and Japan in the case of tea; the Horn of Africa in the case of coffee; and Mexico for chocolate. Early European visitors to China were told how tea was sometimes drunk with a little added milk and, occasionally, with a small amount of sugar when 'this Liquor proves bitter to the taste'. Sugar, though, was marginal and unimportant to Chinese tea-drinkers.[2]

That changed when tea and coffee took hold in Europe and

North America as the popular drinks of millions of people. In the case of tea, the graphs of tea (imported to Europe from China) and of sugar (imported from the Caribbean) matched each other, marching virtually in step from the late seventeenth century onwards. The result was one of the most extraordinary social and cultural formulae we can imagine – tea shipped 10,000 miles was blended with sugar that had been shipped 5,000 miles. And that sugar had been cultivated by Africans who themselves had been shipped across the Atlantic against their will. Behind the humble cup of sweet tea there lay a remarkable global trade – a worldwide transfer of goods, commodities and peoples (with all the necessary commercial underpinnings of finance, insurance and commerce) – which brought together regions and peoples of far-flung corners of the globe. The purpose was to please and to satisfy the tastes of Europeans and their emigrant offspring who settled distant colonies.

Tea, coffee and chocolate appeared in Britain at much the same time, in the mid-seventeenth century, and each was scrutinised and evaluated by contemporary men of science, all part of that remarkable wave of curiosity about exotic items from the far reaches of the world. Flora and fauna, food and drinks, animals – humans even – all fed into the Western world's scientific (and commercial) coming-to-terms with the world at large. Here was an example of the early efforts to transplant and relocate people and plants to different parts of the world, to see if those transplanted entities could be cultivated commercially in regions newly settled by Europeans. This was the story of sugar, coffee and chocolate and, later, tea, when it was transplanted from China to India. Beginning with royal and aristocratic society, it quickly became apparent that tea, along with sugar, could be sold to elite circles, whatever science, especially medical science, had to say about tea. And the prevalent thinking about the properties of tea varied greatly: is it good or bad for health? Can it be used as a medicine for particular ailments? Tea-drinking was a social fashion which gave tea its crucial lift-off.

The Dutch led the way, and their long-distant merchants sold tea to British merchants, who made it available in England. Tea was sold in London by the 1650s; Samuel Pepys first drank it in 1660, and his wife took it as a medicine in 1667.

Pepys also began to drink coffee regularly during that period. The coffee shops which began to spring up in London around that time often sold tea alongside coffee. Unlike coffee, which was relatively cheap from its early days, tea remained expensive. When the East India Company sent some of its tea to the King as a gift in 1664, the company knew that he would recognize the tea as a special luxury. It remained a luxury item throughout the seventeenth century, and consequently remained the preserve of only the wealthy or prosperous elites.

Tea services were also imported from China, and formed the essential and generally beautiful and refined equipment for serving tea, but they were also expensive. By c.1700, London and the emerging spa towns boasted a number of 'tea shops' – some of them run by women. Both there, and in private homes, tea remained a fashionable and costly item, not helped by heavy import duties, a fact which prompted the rapid growth of a new industry which thrived throughout the eighteenth century: smuggling tea to avoid the import duties.

Curiously, to modern eyes, throughout the late seventeenth century tea-drinking was more widespread and popular in the Netherlands than in Britain, largely because the Dutch (unlike the English) had direct trading links and agreements with

China. In Britain, coffee, not tea, had been first to establish itself as a popular drink. As coffee shops proliferated across London, they provided men with a meeting place that lacked the raucous booziness of the alehouse. As a result, they became an important location for a number of social, economic and political gatherings. The coffee shop was a place where men smoked pipes of tobacco, itself recently established as a popular and commercially viable product imported from plantations in Virginia and Maryland. But they were also a place for politics, for the conduct of business, or simply somewhere to discuss news, from home and abroad, via newspapers and prints which became an important feature of the coffee house. Customers heaped spoonfuls of sugar into their coffee to sweeten its naturally bitter taste.

From the start, coffee-drinking was a communal activity, an occasion for male sociability and companionship. Tea-drinking, on the other hand, remained much more domestic and private (and costly), to be enjoyed alone or with a small group of friends or family gathered round a table in the home. John Locke was one who disliked the coffee shop for all the reasons others loved it; he disliked its garrulous sociability, and wanted instead to be alone with a hot drink. He asked friends to send tea from Holland, which he then brewed to entertain his guests. Better still, he sat alone, with his books and papers, savouring a solitary cup of tea.[3]

The transformation in British tea-drinking took place after 1704 when the restructured United East India Company opened up direct trade to China. There followed a rapid and astonishing growth of trade to China. Before 1700, perhaps a grand total of 150,000lb of tea had been imported from China, but in the next five years, more than 200,000lb were shipped.[4]

Tea, only recently a costly drink, now began to become ever more commonplace, praised left and right by scientists, medical experts and by fashionable commentators. Social life now revolved around tea-drinking rituals. In time, tea even found a place in working life, with the growth of the familiar 'tea breaks'. British people, high and low, came to love tea; they consumed it in huge and growing volumes and they wrote lovingly about it.

Shiploads of tea from China landed in London, the tea-chests freighted with Chinese porcelain, and the leaf tea itself proving to be a perfect 'packing' material to safeguard the transport of delicate items on their long voyages from Asia. To avoid theft, the tea chests were quickly moved to dockside warehouses, some of which could store 650,000 crates. By 1767, there were some 7 million pounds of tea in storage; by the 1820s, that had risen to 50 million pounds. Tea had become such a massive business that thousands of London's labourers were employed simply shifting tea between ships, warehouses and merchants around the country.[5]

Quite apart from the different quality of tea on offer, there was a great variety in the types of tea. From the highest, most fashionable and costly brands, down to the roughest of leaves – the fag end of the entire system – tea not only transformed the face of London's dockside labour force, it utterly changed the habits of British people everywhere.

Tea-drinking had slipped its moorings among the upper classes and found a new home among the poorest of common people. By c. 1800, even the very poor had come to regard sweet tea as one of life's essentials. Social investigators studying the diets and the finances of the nation's poor were perplexed to discover that even the most wretched of the poor expected two commodities – tea and sugar. This was to be a recurring theme in studies of the poor from that day to this. Sugar, once a luxury for the wealthy, had become a necessity for the poor. But how had that happened?

It did not stem from improved living standards. In fact, the poorer the individual or family, the more resolute was their attachment to sugar. This taste for sugar seems to have been driven especially by two factors. The first was the sheer volume of imported tea and sugar, which drove down the price of both, helped by a massive smuggling industry that thrived until the duties on tea were finally slashed in 1784. What had, say, in 1700 been accessible only to the rich had, a century later, become available to the poor and was sold in local shops for a matter of pence.

The second factor seems to have been driven by domestic servants, who formed one of the largest single occupational groups at the time; they acquired the habit for sweetened hot drinks from their employers and friends. Men and women who served at table, who worked below stairs, who prepared meals and who served the tea, those who filled the sugar bowls, senior domestic servants who ordered and managed the kitchen accounts, all were among the first working people to encounter tea and sugar. Allowances for tea and sugar began to replace the ration for beer traditionally granted to servants and, gradually, sweet tea replaced beer as part of their food allowance.[6] Household accounts reveal which servants were allocated tea and sugar along with the meals and food provided for their daily sustenance. Sweet tea, twice a day, had become the norm in the servants' quarters by the mid-eighteenth century. Servants also simply took both tea and sugar, tasted it, and liked it, behind their employers' back. Why else did fashionable ladies keep their tea secure in a locked chest? Temptation was kept under lock and key to prevent servants sampling the goods for themselves. Social satires, poems, plays, paintings – all provide sarcastic glimpses into this world. The tea-drinking rituals of the better-off – the styles, fashions, pretensions – were both idealised and then mocked.

So, too, was the way these habits slipped down the social scale – servants aping their betters – to be followed later by other working people following similar routines. It was, after a fashion, the democratisation of culture. Over the course of a century, sweet tea thus became the habit of the common people. It spread not simply from rich to poor, but from town house and rural retreat across the entire country. No longer urban or privileged, the consumption of sweet tea became a national pastime. As Engels was to note in the 1840s, 'Where no tea is used, the bitterest poverty reigns.'[7] All this may seem so obvious because the story is recognizably true to this day.

There were, it is true, great varieties of tea-drinking: weak tea versus strong, one brand of tea rather than another. The overall pattern, however was clear; the British people *needed* their tea – and they needed it sweetened with sugar. They learned how to mix it, how to get the most out of the leaves they bought, often even reusing them, time and again, until all colour and taste had been leached from them. It became a national addiction that raised a really puzzling issue: why had a nation become so attached to consuming two commodities produced at the far ends of the world? Wasn't it strange, in the world where travel and transport took months, that the British (and Europeans at large) demanded goods which their forebears had managed without and, indeed, had not even known about? In 1800, as in 1900, it was asked, 'What could be more British than a sweet cup of tea?'

All was made possible by the Honourable East India Company, founded in 1660, after the Dutch, but soon becoming the most powerful of all European companies trading to Asia. It thrived on tea from China. Having shipped some 20,000 pounds of tea in 1700, sixty years later it transported 5 million pounds of tea. (The Government suspected that as much again was smuggled in to avoid duties.) The Dutch East India Company's tea imports peaked in 1785 at 3.5 million pounds.[8] By the end of the eighteenth century, an estimated 20 million pounds of tea had been imported legally into Britain.

By any reckoning, these are astonishing figures, but they need to be harnessed to the comparative data for sugar. In the words of Sidney Mintz, the success of tea 'was also the success of sugar'.[9] It was a story not merely about European consumers, but about the complex links between Europeans and their various colonies and distant trading outposts. Hostile observers failed to notice that sweet, warm tea gave a feeling of enhanced well-being, although it lacked the nutrition, say, of the traditional beers. To make this formula work, tea-drinkers needed Africans to grow their sugar.

To sweeten their drinks and foods, Europeans were importing sugar on a staggering scale. In 1600, Brazil was the only American exporter of sugar. Fifty years later, Barbados exported 7,000 tons. By 1700, ten colonies in the Americas exported 60,000 tons of sugar, half of it from the Caribbean. Yet within a lifetime, even this astonishing figure was surpassed. In 1750, 150,000 tons of sugar left the slave colonies. On the eve of the American War in 1776, it stood at 200,000 tons, 90 per cent of it coming from the Caribbean.[10]

Not all of this was to be mixed with hot drinks, of course. Much went into the transformed diet of Europeans – into desserts, breads, porridge, puddings. But sweetest of all was tea. What became the most defining, most *British* of concoctions – sweet tea – was a result of Europe's entanglement with distant societies and distant peoples. Sugar and tea had transformed the physical face of distant colonies and countries, and they also created one of the most characteristic social habits of the British people.

Tea-drinking, however, attracted fierce criticism from a string of eminent commentators and writers, from Jonas Hanway to William Cobbett. Some thought it bad for health, some viewed it as a diversion of scarce resources, while others dismissed it as a pointless luxury – the poor seemed merely to be aping the habits of their superiors. More perceptive critics however appreciated the importance of tea and sugar in enhancing Britain's domestic and global trade and power. It underlined Britain's greatness through its trade to Asia, its colonies in the Americas and, as a critical link tying everything together, it confirmed the nation's maritime strength. This was all reinforced, of course, by unparalleled military power at sea.

Along with most of Western Europe, the British had initially turned to coffee, not tea. Native to East Africa and the Arabian Peninsula, coffee-drinking had long been at home in a number of Islamic communities. Coffee shops, a feature of Islamic cities from the Yemen to Algeria, from Iraq to Istanbul, were venues for male sociability, for business discussion and relaxation. Trade and travel between Western Europe, Turkey and Egypt brought coffee, and coffee shops, to Europe. Venice had its first coffee shop in 1629 and others quickly followed in Europe's major port cities. Amsterdam not only had a growing number of coffee shops, but had access to plentiful supplies of sugar from the city's refineries which processed the sugar shipped from Brazil and, later, from Dutch Caribbean colonies. Like tea, sugar and tobacco, coffee also entered Western Europe via apothecary shops, but whatever its alleged medical virtues, coffee established its niche – as it had in Istanbul – as a personal and social pleasure for menfolk gathering at their favourite coffee shop.

The British were quick to switch from coffee to tea. In 1700, they consumed ten times as much coffee as tea. Twenty years later, that began to change. As volumes of imported tea increased, the price fell and the popularity of tea-drinking took off. Coffee maintained its niche, of course, notably in the proliferation of coffee shops. By 1662, there were eighty-two of them in London; and around 550 in 1740. By then, they had become 'a chief focus of social life'.[11] Some catered for the highest of high society, others for the lower echelons. Some effectively became offices (for insurance and banking, for example), but all offered good company and conversation – without the drunken excesses associated with the alehouse. The coffee was served black, but with sugar always on hand to spoon into the drink to counteract the bitterness.

The French, as we have seen, were noted for the amounts of sugar they heaped into their coffee. Parisian coffee shops had emerged, not from the commercial milieu we see in Amsterdam, London and Boston, but from the Ottoman Embassy, liberally dispensing coffee at social and diplomatic gatherings. Lacking the direct commercial links to coffee-producing regions, and without the essential commercial groups to encourage coffee-drinking, Parisian coffee shops struggled at first. They took hold as a location for fashionable aristocratic society, and were unusual in that they sold alcohol alongside coffee; they were, as they remain to this day – cafés, not coffee shops.[12] Even so, sweet coffee was soon to be found in all corners of society. It was served in royal palaces, and sold by itinerant hawkers on the streets of Paris.[13]

The three major Western European cities where coffee shops proliferated from the mid-seventeenth century – Amsterdam, Paris and London – were all linked to a burgeoning sugar trade. In the course of the eighteenth century, French Caribbean colonies disgorged ever-growing volumes of slave-grown sugar, culminating in the massive production in St Domingue in the mid- and late eighteenth century. Amsterdam had established its own links to the sugar-producing regions of the Americas, first in Brazil (which the Dutch had governed, briefly, until 1654). London, meanwhile, was the commercial and financial engine behind the development of Britain's sugar-producing colonies. The end result was that sugar was everywhere by 1700, sweetening the unpalatable drinks which had become an inescapable feature of Western life.

Coffee had initially been imported to Europe via complex trade routes from its native regions. Britain's early coffee came via the Levant, much of it from Yemen, but by 1720 coffee was arriving via the East India Company, although a great deal of it was re-exported to Holland. Europeans were keen, however, to establish their own coffee-producing territories. Indeed, all the major European colonial powers were anxious to establish commercial ventures in their colonies and trading posts, and all of them actively experimented with crops transplanted from one region to another: sugar from the Mediterranean to the Americas; tobacco from the Americas to Europe; coffee from Mocha to Java, and to the Blue Mountains of Jamaica, and high in the spiny ridges of St Domingue; later, breadfruit from the South Pacific to the Caribbean; and tea from China to India. All this activity created 'the Colombian exchange', with peoples, animals and plants being uprooted and resettled far from their native regions, first of all to see if they could simply survive, then flourish, and so become the basis of profitable commerce.

So it was that bitter, black coffee found its place in Western life, first in Europe's main port cities and capitals, later across the face of society, before moving across the Atlantic where it followed a similar route into the new towns and settlements of the Americas. Boston, a large, lively town by the late seventeenth century, had its own coffee shops, modelled on the London prototypes, to provide refreshment and commercial opportunities. They, too, offered that vital mix of coffee and of print, and were soon followed by other coffee shops in New York, Philadelphia and Charleston. Here, too, men found an ideal rendezvous for political and commercial debate. They also became a focus for the rising tide of American anger against British colonial rule and regulation. Throughout the 1770s, coffee shops were associated with American dissent and resistance to British rule, although one in New York provided a regular meeting place for British troops.[14] The seismic American upheaval in 1776, and the real and symbolic throwing of 300 tea-chests into Boston harbour in December 1773, led to a conscious American effort to abandon the British fashion for tea-drinking. It was part of the broader rebellion against what Americans viewed as iniquitous taxation (on tea, in this case) by their colonial masters. Thereafter, Americans turned their back on tea-drinking and became a nation of coffee-drinkers. But even this new nation of coffee-drinkers laced their dark, bitter beverage with sugar.

Today, it is simply assumed that Americans are coffee-drinkers, and although they came to the habit later than

Europeans, their desire for the addition of sugary sweetness was just as compelling. Thus, on both sides of the Atlantic, the development of hot beverages saw a huge increase in the personal consumption of sugar, which had become the natural partner of tea and coffee.

Coffee in North America, though, remained costly and, in 1783, Americans per capita consumed only tiny amounts of coffee each year.[15] Even after independence, coffee remained exclusive and pricey for some time, mainly because of the years of revolutionary upheaval in the Caribbean. (Coffee production in Haiti effectively ceased.) But the return of peace, and the economic development of the early republic, saw a massive growth in American coffee consumption. By 1830, Americans were drinking six times as much coffee as tea; by 1860, nine times as much.[16]

Coffee poured into the new USA. In 1791, less than 1 million pounds of coffee were imported; five years later, it had reached 62 million pounds.[17] When the tax on coffee was removed after 1832, imports shot upwards – 150 million pounds were imported by 1844. By then, the average American was consuming more than six pounds of coffee annually, drinking it at mealtimes and as an occasional drink.[18]

Initially, the American demand for coffee forced the price upwards, but the rapid expansion of coffee cultivation, especially in Brazil, Java and Sumatra, led to a substantial fall in prices. In 1823, coffee cost 30 cents a pound in the USA but, by 1830, it had fallen to 8 cents. When the tax on imported coffee was removed in 1832, the country was flooded with coffee and, by the 1850s, Americans each consumed five pounds of coffee a year. By the end of the nineteenth century, that figure had reached eight pounds. Coffee consumption had outstripped tea by 1830, and Americans were firmly established as a nation of coffee-drinkers.[19]

For all the revulsion against British tea as a stimulant to American coffee-drinking, it was the relative proximity of coffee growing in the Caribbean and Brazil that helped swing North America behind coffee. The development of trade between North and South America saw coffee imported in return for timber (and also for shipping African slaves to Brazil). US coffee consumption was also helped by the large-scale immigration of North European coffee-drinkers. Unlike Europe, however, North America did not have a café culture, and Americans enjoyed their coffee at home, buying green, unroasted beans, before roasting them and turning them into drinkable coffee at home. The end result was that coffee was marketed and sold to housewives, with all the subsequent efforts to create brand loyalties among customers.

Better roasted coffee (as opposed to green coffee beans sold for roasting) emerged from the new coffee technology as the century advanced. Better coffee roasters, grinders and a wide range of new, improved coffee pots, all helped to facilitate the production of better coffee for the American drinker. In 1852, New York had its own Coffee Exchange, and the modern era was ushered in of coffee as a global commodity, bought and traded on the exchange floor, much like cotton, for the rest of the nineteenth century. Increasingly, the US federal government also intervened to ensure standardisation and quality of the coffees sold on the American market.

The major drive in the US coffee market was in packaging and marketing, but green coffee beans remained dominant until c.1900. Thereafter, vacuum packing of roasted and ground coffee beans ushered in a new era of coffee-making and drinking. All this created a massive increase in US coffee-drinking. Between 1880 and 1920, consumption doubled to 16 pounds per head. Over the course of the nineteenth century, imports of coffee into the USA increased ninety-fold.[20]

In the twentieth century, American coffee-drinking became a social activity, much as it had been in Europe in the eighteenth century and in Arabia long before. People, especially clerical workers, took 'coffee breaks' at work and restaurants attracted customers by offering cheap coffee and free refills. Grocery stores sometimes sold coffee as a 'loss-leader' to lure customers to buy their goods. By the mid-twentieth century, coffee was widely viewed as a necessity in American life. It punctuated the working day, it provided a break for women tied to domestic chores, and it seemed to restore energy and alertness to working people when they felt themselves flagging. Throughout the Second World War, it was a basic constituent of military rations. Indeed, that war confirmed coffee's essential role in the US armed services, especially when 'instant coffee' could be transported to all theatres of war and easily concocted by the mere addition of hot water.[21]

* * *

What had happened in the USA in the years between the Civil War and the Second World War was a repeat of the pattern we have seen in Britain a century before. Coffee, like tea in Britain before it, became an indispensable part of everyday American life. It was no accident that soldiers in all America's major wars were given coffee as part of their daily rations (coffee had replaced rum as an official drink of the US Army in 1832). In the Civil War, the Confederate Army reprinted a pamphlet by Florence Nightingale, which included a prescription: 'Coffee for One Hundred Men, One Pint Each.'[22] But severe wartime shortages of both sugar and coffee during the Civil War forced people to resort to a range of substitutes – plants, beans, grains – whatever seemed to provide a dark, coffee-like end product.[23]

In the USA, coffee and sugar went hand in hand as much as tea and sugar were companions in Britain. As coffee consumption boomed, so, too, did the demand for sugar. But the American addiction to coffee cannot alone explain the extraordinary explosion in the US demand for sugar. The massive growth in population in the early nineteenth century, primarily via immigration, created vast numbers of working people engaged in physically demanding labour. These were the consumers of sugar on a major scale. In the thirty years before the Civil War, the rise in personal income (and a fall in sugar prices) produced a remarkable expansion of sugar consumption. In 1837, they consumed 161 million pounds of sugar but, by 1854, that had increased to 900 million pounds. The per capita consumption of sugar stood at 13lb in 1831; thirty years later, that had risen to 30lb.[24] That had more than doubled again by 1900, to 65lb per capita – before peaking in 1930 at 110lb.[25] As Americans imported and consumed ever more sugar, some began to ask: wasn't it time for the USA to cultivate its own sugar? Or perhaps even acquire sugar-growing colonies?

* * *

By the time of the American Revolution, coffee had become ubiquitous in the Western world. Although it had been dislodged by tea as the national drink in Britain, it maintained its distinctive position in coffee shops as cause and occasion of male conviviality, business and conversation. It was served in elaborate rituals with accompanying equipment (some made by the finest of Europe's craftsmen in porcelain) in the highest of high societies, from one royal court to another, and it also helped sustain working people en route to work. Whatever the location – a City of London coffee shop humming with affairs of trade insurance and overseas commerce, a fashionable Parisian café, the Palace of Versailles or in the argumentative political crucible of a Boston coffee shop – the bitterness of black coffee was tempered by sugar. Tea had conquered the British home, from the highest to the lowest. It was organised, prepared and served by the women of the house. In more fashionable homes, at teatime and at the start and the end of the working day among humbler folk, tea was part of British domestic and family life. Coffee was both more public and more masculine, a focus for men's social and commercial activities. Yet wherever tea or coffee were served, in a palace or a hovel or in public coffee shops, there, too, we find the sugar bowl. Hot drinks were always sweetened. So, too, were a remarkable number of Western foodstuffs; sugar had become an indispensable ingredient in food as well as drink.

7

Pandering to the Palate

Pick up practically any modern cookbook, and you will find sugar listed among the ingredients required in a modern kitchen. Sugar is a basic ingredient in the supplies of any self-respecting cook. Long before the Western world discovered its taste for sugar, the cuisines of many distant societies had made elaborate use of cane sugar. Though the arrival of sugar in Europe had been symbolised by elaborate displays of sugary confections and sculptures, all this had remained within the world of the rich and influential. Less privileged members of society had to acquire their sweetness from traditional, cheaper sources, most notably honey. But the expansion of sugar production in the Americas in the early seventeenth century enabled sugar to slip its moorings from the world of privilege and to find favour among all strata of society. Sugar followed the pattern of other exotic commodities which had also once been the prerogative of Europe's elites and those who were excessively wealthy but didn't have the corresponding status, such as the traders who made their money in commerce but who lacked superior social standing. During the seventeenth century, sugar settled among the common people, in town and country, and they used it both in their drink and their food.

Sugar quickly entered the cuisine and the diet of Europe at large. French cuisine adopted sugar not simply because it was recommended by French doctors in the seventeenth century, but because of the way it obviously enhanced certain foods. Adding sugar to oats, for example, made them fashionable in France, where they had previously been plebeian. Working people across Western Europe began to add sugar to basic foods and drinks – French peasants and workers in Gdansk both enjoyed sweet coffee; sailors' wives in Calais drank sweet tea. Polish workers liked adding sugar to their food, and the habit quickly took hold in North America. John Winthrop noted in 1662 that Indian corn was made tasty by the addition of sugar.

In a number of very different societies, the Catholic Church was instrumental in encouraging a taste for sugary items among the poor. In Mexico and Goa, the Philippines and Mozambique, nuns (in keeping with an older Islamic tradition) made sugary delicacies to sell to worshippers, often based around a religious theme.[1] It was the French, however, who effectively introduced sugar into Western cooking via their pioneering and perfecting of sweet desserts, although, again, they were following a well-trodden Islamic culinary path. The fact that we sometimes call desserts 'sweets' is a giveaway. It was, again, in the seventeenth century that desserts emerged as sugary concoctions to complete an elaborate meal. Before then, there had been no real distinction between sweet and savoury dishes, though many dishes contained a sweet ingredient. All that changed in the course of the seventeenth century when sugar came to be concentrated in desserts.

No one can really explain *why* the distinction emerged, why sweet desserts became a distinct and culminating feature of French meals. But thereafter, a sharp divide developed between sweet and savoury, a distinction which was to become a striking feature of Western cuisine everywhere. The timing, however, was no accident. This change in French cuisine evolved in the years when cane sugar became more commonplace, thanks to the development of the French Caribbean sugar plantations. By 1700, wealthy French people began the day with sugar, eating a breakfast that consisted of drinking chocolate mixed with sugar, along with bread or brioche.[2]

Elaborate, formal French meals and menus evolved as an occasion in three parts, with the last course, dessert, generally cold, sweet and, in fashionable circles, sometimes remarkably ornate. This distinctively French version of dinner was greatly influential across Europe, largely because, by the eighteenth century, France had become the dominant cultural force in Europe. Even the Russian court spoke French, encouraged largely by the admiration for French culture by Peter the Great and Catherine II. Wherever the French style of dessert took hold, the dishes created used lavish additions of sugar – ice-creams, charlottes, jellies, parfits, sundaes, cakes, pies, syllabubs – all and more relied on sugar. The English versions tended to be altogether more modest, though, in imitation of the French, dessert came last. According to Chamber's *Cyclopaedia* of 1741, dessert consisted of 'fruits, pastry-works, confections, etc.'.[3]

This culture of elaborate cuisine, at which the French remain the masters, was still founded, of course, in the world of privilege, and it was reflected in the development of printed cookbooks. Today, cookbooks of all sorts and varieties fill the shelves. There are even bookshops specialising in nothing else but cookbooks. It is a story which begins, in its modern form, in the mid-seventeenth century. Between 1651 and 1778, some 230 cookbooks were published in French alone, although the pattern was quickly followed in other Western countries. New in Europe, cookbooks belonged to a publishing (and cultural) tradition which stretched back at least to the early days of Islam. Now, in early modern Europe, and slightly later in America, such volumes were aimed at a widening readership. Cookbooks were designed primarily for literate females who not only managed the kitchen and the household staff, but who sometimes needed guidance and instruction. Those books were also a reflection of a more fundamental shift taking place – the habits of society's elites were being shared by new social groups. People whose newly acquired affluence, from trade and commerce, their wealth sometimes even surpassing the riches of their royal and aristocratic superiors, were keen to share the pleasures and luxuries of their social betters. Like clothing, houses, carriages and social manners in general, food offered a good way of emulating the lives of people at the top of the social order.

In both France and Britain, the most notable group keen to make their social mark were the Nabobs from India and the sugar planters from the Americas. Costly lavish homes, extravagant parties, spending on a scale which raised even aristocratic eyebrows, all came to define the sugar barons – the plantocracy – the name itself a blend of planter and aristocracy. They had the money, and they wanted the social cachet that went with it. The sugar barons had made their money from sugar and, as if to prove it, sugary desserts adorned their dining tables.

All this belonged to the world of wealthy people, but as sugar became cheaper and more widespread, it naturally found its way into food across the social scale. The middle social orders sought to adopt the sweet-eating habits of those above them and, by the early eighteenth century, puddings (which came in a great variety of tastes, shapes and sizes – but always sweet) had become a favourite dessert even in modestly prosperous homes. Indeed, the word itself – pudding (like 'sweet') – became a British term for dessert.

The relationship between the English and their puddings was a well-established cultural landmark by the late eighteenth century and was used, for example, to telling effect by late-century caricaturists. The increasing number of cookbooks all sought to instruct the nation's wives and daughters in the ways of kitchen management and cooking, and all inevitably turned to the blessings of sugar.[4] What slowly emerged was a codification of rules and conventions both for cooking and for serving meals, a codification which reflected the formality of the French culinary culture from which it had originally emerged.

By the mid-eighteenth century, however, French cookbooks had shifted their social focus. In 1746, *La Cuisinière Bourgeoise* established a new template for cooking. As the title makes clear, the book was aimed at a much lower social class than royalty or the aristocracy, which had been the traditional market for earlier cookbooks. One major problem facing such books was the assumption that there would always be plenty of food – and plenty of money to indulge in elaborate cuisine. But at critical points in the eighteenth century, France was plagued by serious food shortages and real famine.[5] It reached its famous crisis in the Revolution of 1789.

Notwithstanding the recurring problem of hunger, sugar found a secure home in French cuisine in the two centuries before the Revolution. But what impact did it make on the lower orders? What possible use could elaborate desserts, even sweet puddings, be to humble labourers – rural or urban – whose main task in life was simply to earn enough to eat? Domestic servants might encounter sugary delights via their workplace and the kitchens and dining rooms, but what about the rest?

The plebeian diet was, at best, poor. Low earnings left little spare for luxuries, even in years when the standards of living were rising a little. Yet sugar was obviously a luxury. People had survived without it for centuries, and any commodity shipped from the Caribbean arrived loaded with meaning, the very definition of a luxury item. Critics in both France and Britain thought it absurd that the poor, people with few resources, should yearn for sugar. In the British case, the attacks on sugar were directed largely against sugar in tea. Time and again, commentators railed against the consumption of sweet tea – people would be better advised to spend their money on basics such as bread, for example. Yet despite such frequent criticisms, the lower orders turned, in growing numbers, to what many continued to view as a luxury – notably tea and coffee laced with sugar.

Such criticism of sugar rang like a refrain throughout the eighteenth century on both sides of the Channel, as critics denounced the love of luxury among the common people. More perceptive authors recognized, however, that sugar had already become one of life's necessities – it enhanced people's lives, strengthened them for their daily tasks, and added some kind of comfort to a miserable existence. In any case, the kind of sugar consumed by the poor – like the tea they drank – was always the very poorest and cheapest sort. Theirs was not the costly, top-end sugars of the fashionable households, but the literal scrapings of the poorest sugar loaves, which they added to the cheapest of oats.[6]

As the Caribbean plantations disgorged ever more sugar (and other tropical foodstuffs), sugar prices plummeted by half between 1630 and 1680. Sugar imports to Britain doubled, then doubled again. Sugar consumption doubled in the forty years to 1740, before doubling again by 1775. In a little more than a century, the sugar consumed by people in England and Wales increased sixty-fold, while the population scarcely doubled in the same period. The per capita consumption of sugar in 1700 was 4lb, rising to 8lb in 1729, 12lb in 1789 and 18lb by 1809. Sugar was now being used extensively – and not simply in tea. It was added to a range of basic foodstuffs – wheat, oats and rice became significantly more palatable. Not unlike the ancient habits of sweetening foul-tasting medicines, the poor made their simple, bland diet tastier by adding sugar. Sugar also became part of the diet of poor people via the by-products of sugar manufacture – molasses and rum. It augured ill for the future, although no one realised it at the time.

The diet of poorer people was augmented by spreading molasses on bread, thus converting a tasteless morsel into an acceptable dish. Sugar and molasses added a pleasant taste to what had long been meagre meals: sweet tea and coffee, bread or porridge for breakfast; a lunch of fried potatoes; and a similar dish for supper, again with bread or oatmeal, and weak, sugared tea or coffee. For people whose working lives were arduous and protracted, these additions of sugar to food and drink refreshed and revived – and provided essential energy.

Even so, it was a meagre, minimalist diet which regularly shocked curious outsiders and social investigators, who became a permanent feature of labouring life from the mid-eighteenth century right down to the present day. In the nineteenth century, the poor augmented their meagre diet by the addition of sugar-filled jams.

Beneath these generalities there lay a revealing fact, albeit one which had ancient origins. Foods were shared unevenly within working families. The best bits – when available – were reserved for the main breadwinner (the phrase itself revealing). And that normally was the man of the house, although this was to change dramatically in the nineteenth century with the rise of modern textile industries with their preponderance of female industrial workers. The breadwinner needed the physical strength to work at arduous tasks and to bring home a weekly wage. That often left wives and children with the scraps, and with the cheaper or leftover items, and it was they who consumed most of the family's sugar. Yet in the textiles industries – which formed the engine of Britain's industrial revolution – women and young children *also* worked long, energy-sapping hours. Survey after survey, especially in the nineteenth century, revealed the importance of sugar: 'Factory women survived on bread, sugar and fat, supplemented by portions of meat . . .'[7] Jams, heavily dependent on sugar, and later treacle (dispensed by sugar refineries direct from their vats into jugs brought to the refineries by working people) were spread on bread, and became a basic sweet ingredient in plebeian diets across the face of Britain. Sugar thus became a vital source of sustenance and energy for a growing population of urban industrial people; it was a key ingredient in their meagre diet and a necessity in their hot drinks.

Sugar also transformed the way people preserved their foods. Previously, people had preserved fruit in honey and various syrups. At first, sugar enabled apothecaries – later, the confectioner – to create preservatives by boiling fruit and other ingredients, mainly for medicinal purposes. (The process was utterly transformed in the early nineteenth century by the invention of a new bottling process, later by canning.) Whatever the process, whether at home or in new bottling and canning factories, sugar was the essential ingredient, and with the advance of the modern food industry, sugar was added in very large volumes, and this was true for a range of foodstuffs. What made this process possible on an industrial scale was not simply the emergence of the science and the technology of treatment and bottling, but the availability of cheap sugar in huge industrial quantities. Sugar was everywhere. In its various guises, sugar had become an inescapable feature of everyday life. It was like tobacco, both a source of strength and a consolation – 'the general solace of all classes' – but especially of working people.[8]

Sugar had entered people's lives both at home and at work, and it was part of the routines of work itself. Bread, spread with sweet jam (or simply speckled with refined sugar) was carried to work, to be eaten at lunchtime or, increasingly, at the 'tea break'. People whose lives were increasingly regimented by machines, and by the new industrial disciplines that emerged from mechanised work, were granted a break from the arduous monotony, and it was in those brief respites from the demands of the machine that they drank sweet tea and coffee and ate a sweet snack carried from home. Sweetness at the breaks made strenuous work in factories more tolerable, and provided some of the necessary energy required for the tasks ahead. Sugar, not long ago a luxury for the rich, thus became a necessity for working people.

Even so, many critics continued to regard the transportation of food and drink vast distances a strange way of feeding the population, especially the poor. Nonetheless, sugar (and molasses and rum) continued to make their mark among working people. In colonial North America – which was to become a land flowing with material abundance – settlers were greatly reliant on imported goods. Until the local economies matured, and until both land and people began to yield the abundance which came to be associated with that vast continent, colonial North America needed a great range of imports. Americans enjoyed some things in abundance, of course – there were plentiful fuel supplies for cooking and heating, copious running water (for grinding) and a profusion of land for cultivation and for animal grazing.

Yet colonial Americans imported huge volumes of foodstuffs. They also consumed huge volumes of the slave-grown produce from the Caribbean colonies to the south. In large part, this was organised by design; the British tied the northern colonies into the wider imperial system. The slave islands needed American commodities to maintain the sugar plantations. And, in return, they shipped sugar, molasses and rum northwards. More than 75 per cent of the value of all items imported into North America from the Caribbean in the late eighteenth century consisted of sugar, rum and molasses. As a result, rum was *the* favoured drink given to labourers in North America, much as beer was in Britain. In 1770, an estimated 7.5 million gallons of rum were shipped to the northern colonies. There, men and women, free and enslaved, drank rum on a daily basis. It provided perhaps one quarter of all the calorie needs for an active adult. We also know that, on the eve of the American Revolution, some two thirds of American adults drank tea twice a day.[9] They, like their British counterparts, sweetened it with sugar from the Caribbean.

Behind these patterns of the American diet lay an issue which was to become a source of great American contention, and the cause of political opposition to the British – the taxation on imported foodstuffs. The British state enjoyed income via the duties levied on commodities shipped into the northern colonies. The string of British Acts of Parliament which threatened to increase taxation on such items – the Molasses Act 1733; the Sugar Act 1760; and the Tea Act 1773 – was to create deep American resentment which helped propel Americans towards independence.

Each of those Acts was directly related to the consuming passion for sweetness in food and drink, and they were all, of course, directly related to slave labour. British governments saw those commodities as a source of income, but Americans viewed them as an intrusive hand in matters over which they had no say or influence. Sugar was, at one and the same time, both cause and occasion of North American popular taste, and of rising American antipathy to British rule.

Molasses emerged as a key commodity in North America. From the Massachusetts fishing communities in the north-east to the slaves in the Old South, molasses from the Caribbean was a major ingredient in the American diet. The American poor especially relied on molasses, on bread, as a staple diet. Slaves liked molasses mixed with cornmeal and pork. Molasses was also used in the manufacture of North American alcoholic drinks. Grocers sold molasses from barrels in their shop, and that sticky substance established itself in a range of popular, cheap American recipes, from gingerbread to Boston brown bread – even in Boston baked beans.[10]

When the North American colonies broke away from British control in 1776, they had come to rely on sweetness imported from the Caribbean slave colonies. But by then, so had the people of Europe. On both sides of the Atlantic, sugar and its sweet by-products had insinuated themselves into the diet and drink of millions of people. In the form of rum, it brought solace and comfort to millions, although it was to have disastrous consequences among the native peoples of North America.

8

Rum Makes its Mark

T HE IMPACT OF cane sugar on the modern world went far beyond the urge for sweetness. Extracting refined sugar from the sugar cane involved industrial processes which, at first, began in relatively simply factories on or close to the plantations, and was then completed in sugar refineries in Europe and North America. This sugar industry linked together two distant parts of the world – tropical cultivators and northern industries in temperate countries – which remained a basic feature of sugar production for centuries. That long-distance and protracted international system produced sugar crystals – refined sugar – which people added to their food and drink. But it also created a number of by-products which themselves became important features of modern consumption habits.

The cane cut in the sugar fields was carted into factories to be crushed, boiled, evaporated and then filtered into pots and barrels. Bigger plantations had their own factories, and even at an early date they formed an industrial complex rooted in the rural heart of sugar cultivation. Long before modern factories characterised Europe and North America, sugar factories dotted the landscape of the colonies, belching steam and smoke into the tropical sky, and announcing that the sugar crop was in full swing. Smaller sugar cultivators sent their cane to the nearest plantation factory for processing. Much later, large-scale 'central factories' emerged to undertake the same process.

Converting cane into sugar generated a trail of by-products and waste matter: crushed canes ('*bagasse*', which was later used as fuel); a liquid residue of impurities; and molasses. Molasses could then be distilled again, and further processed to produce rum. Although rum-making had long been familiar in Islamic sugar production, the alcohol – prohibited as a drink in Islam – was used in medicines and perfumes. Europeans, on the other hand, with their own traditions of distilling strong spirits, had no cultural restrictions on drinking alcohol. Brazil produced crude rum in the mid-sixteenth century and sugar planters noted that African slaves enjoyed it. One early critic remarked (in 1648) that it was 'a beverage fit only for slaves and donkeys'.

Providing the crudest rum to the enslaved labour force continued throughout the history of slavery in the Americas, but attitudes to rum changed when the drink began to prove its commercial value. In fact, a variety of alcoholic drinks emerged from the sugar industry. An Englishman told of a drink he encountered in Puerto Rico in 1596 made from molasses and spices, and other forms of fermented alcohol from sugar were reported in a number of slave colonies. Until rum became a viable export commodity, however, many planters were happy to allow slaves to use the residue from sugar manufacture to make their own alcoholic drinks.

By the mid-seventeenth century, rum had become an established export commodity in its own right. The exact origins of rum's commercial production remain uncertain, but it seems likely that it effectively began in Barbados and in Martinique. Dutch refugees, expelled from Brazil, may have helped establish rum distilling in both those islands. By the 1640s, rum was being manufactured in Martinique; a decade later, it was well established in Barbados. The original Barbados rum was described as 'a hot, hellish and terrible liquor' and was known by various names, 'Kill Devil' being perhaps the most revealing.[1] Much of it was consumed on the island (by the 1670s, there were an estimated 100 taverns in Bridgetown), although some was exported to North America and Britain. Rum punch (a staple of the modern Caribbean tourist industry) was popular among sugar planters by the 1660s. A century later, it was common in Europe and North America, hence the appearance of 'punch bowls' in taverns and on fashionable dining tables.[2]

By then, rum was hugely popular in the Caribbean, and visitors were often struck by the levels of drunkenness in the islands, and by the associated levels of ill health. Widespread, heavy rum-drinking in Barbados was often followed by 'dry belly-ache'. It was discovered that the problem was caused not by rum itself, but by lead poisoning, although the culprit was only finally located in 1745 – the lead piping in local rum distilleries.[3] By then, there was a thriving rum export trade to Britain and, more commercially significant, to New England – especially to Rhode Island. Distilleries designed to produce rum were founded in major ports on both sides of the Atlantic. In Bristol, the local copper and brass industries specialised in making equipment destined for the distilleries in the Caribbean.[4]

As sugar cultivation spread throughout the Caribbean in the course of the seventeenth century, rum emerged as a popular drink throughout the Americas and Europe. It was now established as a lucrative tropical product in its own right and, like sugar, it yielded handsome returns to the British state in the form of duties levied on imports. Rum also helped to change the physical appearance of the sugar islands. Not only did the plantations devour the wild landscape and convert it to orderly fields, but the sugar factories and the distilleries used for producing sugar and rum saw the development of industry in the middle of rural activity. The Caribbean landscape was dotted first by hundreds of windmills, then by smoking and steaming chimneys attached to factories and distilleries.

By c.1700, rum distilleries in Barbados had taken on a recognizably modern form, with their custom-made copper stills (hence the term 'coppers'), with large metal vats, and all the necessary piping to contain and process the liquids. Although it was on a small scale at first, as volumes increased, and as taste and demand grew (locally, but especially in North America), rum, like many other exotic products, passed from the apothecary shop to the tavern-keeper. A great deal of rum was consumed in the rum-producing regions by master and man, planter and slave. On Martinique, servants and African slaves each drank, on average, an estimated 3.5 gallons of rum per year. Much the same was true of military staff stationed in the Caribbean. They drank rum for pleasure, of course, but also as a result of prevailing medical advice. Alcohol was seen as a vital medicine in maintaining the body's temperature against the threat of local diseases. The conviction spread that rum was especially important for white troops stationed in the tropics; it was also recommended for men standing guard during the night.

British troops in North America were consuming a *monthly* average of 3.5 gallons of rum – twelve times that of the slave population that produced it.[5] There was, then, a thriving rum trade on each sugar island, and between the islands, although the main impetus behind the development of Caribbean rum was the export trade, particularly to North America.

Rum exports from the Caribbean increased rapidly in the late seventeenth century. In 1664–5, 102,744 gallons of rum were shipped from Barbados; thirty years later, that had risen to more than half a million gallons.[6] By c.1700, rum had become an important source of income to planters throughout the Caribbean. Above all, perhaps, that rum provided comfort and pleasure for untold numbers of drinkers in taverns and 'grog' shops on both sides of the Atlantic. Rum-drinking spread from the plantations to the dockside, and then to all the major ports and their wider hinterlands around the Atlantic. Rum became the mainstay of seamen plying their trade throughout the extensive legs of global maritime commerce. Sailors had traditionally needed alcohol to sustain them through the rigours of long-distance sailing for weeks and months at sea.

After 1731, rum also established itself as the daily sustenance and pleasure of men serving in the Royal Navy. Indeed, it is hard to overestimate the importance of rum to sailors in both the commercial and naval fleets. In fact, the ration on Royal Navy ships continued until 1970. The hardships endured, for instance, by the crew on Captain Cook's epic voyages to Australia and New Zealand in the 1760s and '70s were eased by a daily one pint of rum (a half pint for boys), doled out at noon and 6 p.m.[7]

The most important market for Caribbean rum was North America, where it quickly spread from dockside taverns to the remotest of frontier settlements. It was drunk by 'yeoman farmers, fisherfolk, Chesapeake planters, African slaves, Native Americans, and frontier fur traders . . .'[8] By 1700, Barbados exported a yearly average of almost 600,000 gallons of rum, but only a tiny fraction went to England and Wales; the bulk went to North America, and there was also a lively trade to Spanish America. And although the islands cultivated a wide range of export crops, more than 90 per cent of the trade from the British islands to North America consisted of sugar, rum and molasses.[9] Another receptive market for Caribbean rum was Ireland and, in the process, the Irish became a nation of rum-drinkers.[10]

Rum became much more than a pleasurable popular drink in North America; it was also a means of exchange – a form of currency. The New England merchants importing most of North America's rum lacked any real money supplies, so they bartered for their rum, paying for it with cargoes of North American goods – foodstuffs, fish, timber and pitch – all of which they shipped south to the Caribbean. In this way, the economy of North America helped to clothe, house and feed slaves in the Caribbean.

North America's taste for sweetness in drink and food was established in the years of colonial rule, and it laid the basis for what followed when the American Republic developed into the most powerful economy and country on earth. From the early days of European settlement in the Americas, there had been strong economic and personal links between North America and the Caribbean; they were particularly strong between Barbados and the Carolinas. There were also regular migrations of people from one region to another. As the Caribbean plantation economies evolved, the islands naturally turned to North

America both for markets and for a wide range of necessary goods – timber, for construction and roofing, and foodstuffs (especially cod fish) to feed the Africans. In return, and by way of payment, the northern colonies received Caribbean produce. Above all, they liked the by-products of the Caribbean sugar industry – rum and molasses.

North America also needed sugar, and although there had been early efforts to grow sugar cane in North America, notably in Virginia, American sugar had to await the settlement and American acquisition of Louisiana in the early years of the nineteenth century before local sugar production took off. Until then, Americans imported and consumed sugar from the Caribbean, although it remained costly. As in Europe, the crude sugar shipped to North America underwent further refining in major ports on the east coast. There was a sugar refinery in New York as early as 1689; a century later, there were refineries in seven American cities, although, by then, most of the sugar had passed through refineries in Philadelphia. As in England, the final product, refined sugar, was sometimes sold in what, to modern eyes, seem to be unusual places. Evan Morgan's shop in Philadelphia offered corsets and children's coats alongside 'very good Chocolate, Wine, Rum, Melasses, Sugar . . .'[11]

The social importance of sugar in colonial America followed the pattern already in evidence in Europe. At first it graced the tables of more prosperous Americans. In the major cities, or in the more elaborate country residences (on plantations, for example), people flaunted their standing, as did their European peers, by equipping their homes with elaborate coffee and tea sets – each with the necessary sugar bowl. Thomas Jefferson, for example, experimented with making ice cream, which required lots of sugar. We also know that sugar was used as a basic ingredient in colonial American cooking, because cookbooks recommending sugar as an ingredient had begun to appear in the kitchens of prosperous Americans. In general, however, sugar remained a luxury item in North America much later than it did in Europe. Poorer Americans, on the other hand, sweetened their foodstuffs with honey, maple syrup or molasses, and the cost of sugar kept consumption low. In the early years of the new Republic, for instance, the average American consumed only 8lb of sugar each year. A century later, in the 1890s, that had risen to 80lb a year.[12]

What colonial North Americans seemed to enjoy most of all was Caribbean rum, its initial popularity spreading from sailors who acquired the taste in the Caribbean and at sea. It was also actively encouraged by the Dutch who had a long-standing commercial interest in the rum trade, although the Dutch were eventually excluded from North America by the imposition of British Navigation Acts, designed to keep trade firmly in British hands. It was this policy which, by 1776, became another major reason for America's growing alienation and dislike of British colonial control. These British restrictions on trade led to a flourishing smuggling trade which ensured that rum continued to flow into North America via Dutch traders.

New Englanders also produced rum from cheap, smuggled French molasses, much to the disgruntlement of British Caribbean planters and British officials keen to maintain British economic dominance over the French. Even a tax on imported French molasses did little to prevent New Englanders acquiring supplies for their rum distilleries. The Peace of Paris in 1763 severely hurt the French. Not only did they lose Canada, but their Caribbean islands were henceforth allowed to export *only* molasses and rum. New England merchants made the most of this by shipping ever greater volumes back to the northern colonies. All this prompted closer scrutiny by the British, with the resulting Acts to regulate North American and Caribbean trade, and the imposition of duties on sugar and molasses (regularly revised to suit British interests) and all serving to aggravate North American – and sometimes Caribbean planters' – feeling against the British.

Caribbean molasses was widely used as a source of sweetening in North America, and was also used to produce local beer. But its main use was in making local, North American rum. At the mid-eighteenth century, 2.7 million gallons of rum flowed from distilleries in Massachusetts. It was sold in taverns throughout the colonies (Boston alone had 177 in 1737) and was sold throughout the colonial settlements as far as the distant frontiers, where it sustained and emboldened the military, and lubricated trade with the Native Americans. The North American colonies seemed awash with rum. Indeed, throughout the American colonies (and what was to become Canada), rum was a prized commodity that could be exchanged for most other items.

All this is evidence of the degree to which the sugar economy, in all its complexities, had permeated the Atlantic world by the mid-eighteenth century. People all over the Atlantic region had come to depend on sugar for their drinks and food, but they also enjoyed sugar's derivatives in equal measure. In North America, Caribbean rum had established itself as a major drink, especially among labouring men. It proved an ideal drink to cope with the rigours and the hardships of colonial life. It also became a vital element in the fur trade with Native Indians and, despite efforts to control its impact, it was to have a dire, malignant influence on those communities – harm that was later to be exacerbated by whisky. Although American colonists also established their own alcoholic drinks – wines, and especially beer – those drinks made little headway against rum imported from the Caribbean.

Barbados dominated the early rum trade and, by the end of the seventeenth century, that tiny island was exporting 600,000 gallons of rum annually. Most Barbadian rum went to New England and to the Chesapeake region dominated by slave-grown tobacco. An estimated 250,000–300,000 gallons of rum were shipped to the Chesapeake alone. A similar story could be told of French North America, with French legal restrictions demanding that colonists import and consume only produce from French Caribbean islands. That rum was a key ingredient in the fur trade with Native Americans. French officials, like the British, were torn between the economic advantages created by rum exchanged for furs and pelts, and the obvious damage wrought by rum among Native Americans – and among African slaves. Colonial officials and the occasional local cleric united in arguing that drunkenness was having a corrosive impact throughout the colonies. Despite this being also true of large numbers of European settlers, it seemed especially damaging among the Native American Indians – and the main source of that problem was rum and molasses imported from the sugar islands.[13] And it is from around this time that the phrase 'a bender' was first coined – it was originally a Seneca Indian expression for a bout of excessive drinking.[14]

Wherever Europeans settled and traded in North America, officials complained about levels of drunkenness. They tried to curb it, they railed against it, and they tried to limit the import-ation and sale of rum. But the harsh reality was that the New

England economy *needed* rum; it was the essential means of exchange for a string of local export commodities. Without imported rum, there could be no export of local goods which were so vital to the well-being of the economy of the northern colonies. Boston merchants were understandably anxious to maintain their lucrative trade to the Caribbean. Moreover, the increase in numbers of New England rum distilleries made imports of Caribbean molasses even more important and, despite the regular complaints, rum and molasses, two by-products of slave-grown sugar, had become an integral feature of the economy of both French and British North America. They were also central features in the social lives of settlers, slaves and American Indians alike. As long as that remained the case, local denunciations or prohibitions against rum would remain utterly ineffective.

* * *

By the late eighteenth century, the influence of sugar and rum had spread far beyond the obvious markets in Europe and North America. Because France prohibited the import of rum (to protect its wine and brandy industries), French rum producers were forced to look elsewhere for markets. When Spanish rum took off in the mid-eighteenth century, especially in Cuba (although it remained on a lower level than the French and British industries), it, too, faced hostility from the Spanish wine and brandy industry. This allowed the British and French to export large volumes of their own rum to Spanish colonies in the Caribbean and South America. The French islands of Martinique, Guadeloupe and St Domingue exported huge volumes of rum ('*tafia*') to Spanish colonies.

By then – in what seems a bizarre, ironic twist – molasses and rum had also found their way to West Africa, normally via Europe's major slave-trading ports. Slave-grown tobacco, rum and molasses became a popular export item from Brazil to West Africa, where it was traded for yet more Africans bound for slavery in the Americas.

By the time of American Independence, rum and molasses were important features of the American economy. Both were widely consumed in North America, although they had also spread wherever Europeans settled, traded and colonised. British soldiers also received a regular rum ration. Soldiers and sailors who survived the fearful mortality rates in the Caribbean, all returned home with a taste for rum.

Rum also took hold in the British population at large, promoted by the West Indian planters' lobby. It was seen as an alternative to the disastrous levels of gin consumption that so plagued the country until the restrictions of the Gin Act of 1750. Rum punch seemed more benign and less corrosive than gin, and it had been a popular drink among planters from the early days of the rum industry. By the 1730s, it had established its popularity on both sides of the Atlantic. It was unusual in that it began as a popular drink before gaining a fashionable following in the form of 'punch' in the late eighteenth century and, as we've seen already, the prevalence of expensive 'punch bowls' among the fashionable status symbols of prosperous society.[15] The rum trade ate into the long-established British reliance on continental Europe for wines and brandy and, by 1733, Britain imported 500,000 gallons of rum. Although the sugar lobby tried to woo the wealthy from their love of continental wines and spirits, rum's major, secure market was significantly lower down the social scale.

By the end of the eighteenth century, the British taste for rum was being supplied not solely by Barbados, but by the burgeoning sugar economy of Jamaica. In the early 1770s, 2 million gallons of Jamaican rum were shipped to Britain. Jamaica's sugar plantations were now earning between 15 and 20 per cent of their overall income from rum – although the percentage received by plantations in Barbados was even higher.

The Irish, too, enjoyed their rum. In the 1760s and '70s, Ireland imported almost 1.5 million gallons of rum from Barbados. The main market for Barbadian rum, however, was North America. British Canada also liked it, importing 600,000 gallons of Barbadian rum in 1770. Jamaica (which was less reliant on rum exports) shipped c.900,000 gallons to the northern colonies on the eve of the American war. The Caribbean planters by this time had come to rely on rum for a large part of their livelihood; upwards of 25 per cent of their income came from the sale of rum.[16]

American independence in 1776 posed a threat to the Caribbean sugar economy and planters feared losing vital supplies from North America but, once again, smugglers helped them out – rum headed north and American supplies headed south, shipped through the Danish Caribbean islands. There was a game of fiscal cat and mouse between the British and Americans about supplies to and from the Caribbean. In the event, the real threat to Caribbean rum in North America came from the rise of the local whisky industry, and the American revulsion against all things British. The USA had begun the process of going its own way politically and culturally, and cast aside British habits by choosing to embrace coffee and whisky, rather than tea and rum.

By the time the USA had set itself on the road to secure nationhood, rum and molasses had found a receptive home throughout the Americas. It had also travelled at sea in Britain's huge naval and mercantile armadas, and in military bases dotted around the globe. Grog shops in New England and on the docksides of old England dispensed the drink, while distilleries throughout the Atlantic world converted slave-grown molasses into rum for local consumption.

Rum even found eager customers among pioneering settlers in Botany Bay. Indeed, some of the early complaints of settlers in Australia after 1787 were about the extortionate prices demanded by local traders for rum. Government officials were also accused of making money by trading in rum with the convict settlers – the same settlers who had been given an allowance of sugar to mix with their tea as part of their daily rations on the first convict ships.[17] Even the pioneers clinging precariously to the new experimental colony of Sierra Leone (mainly freed slaves from London and Nova Scotia) were allowed a rum allowance.[18] And everywhere across the Americas, rum was doled out to enslaved Africans to make their labouring lives more tolerable. Rum also began to play an important role in a variety of religious ceremonies among slaves in the Americas – and among Africans in their various homelands.

The irony, again, is clear enough. Slave-grown sugar, and the molasses and rum derived from processing sugar cane, made slave life more tolerable. Slave-produced rum softened the hardships of men on warships and slave ships, in military camps and in times of conflict, and in the precarious settlements founded by Europeans right around the globe, from the American frontier to Botany Bay. It was as if slaves were producing a lubricant to ease the hardships and miseries of their own lives, and those of their oppressors. And everything hinged on the cultivation of sugar.

Rum may have eased the arduous and stressful lives of settlers and labouring people across the northern colonies, but it also played a disastrous role in weakening the resilience of native people. Rum had a devastating impact on the native people of the Caribbean islands. Although fermented alcohol had played important roles in traditional Taino life, rum brought a new, destructive dimension to their drinking habits. The story of the damage wrought by rum was a precursor to what was about to happen among native people in North America. For all the apparent pleasures they brought, by 1700, sugar and rum were already revealing their power to corrupt life in all corners of the Atlantic.

* * *

At the heart of this tangled web of Atlantic trade and commerce lay Caribbean slave-grown sugar. It had spawned, in rum, a vital source of income, a means of paying for imports to the islands. In the process, it had established massive markets of rum drinkers in America and Europe and it had served – like sugar itself – to fortify and strengthen vast numbers of labouring people on land and at sea. Behind it all lay the acres of sugar cane cultivated by enslaved Africans.

Rum's influence could also be found in some unlikely places. Most surprising, perhaps, was its impact in Africa. Europeans shipped an enormous variety of items to Africa's Atlantic coast in exchange for slaves. Textiles from Asia, cowrie shells from the Indian Ocean, hardware and guns from Europe, French wines, glass beads from Italy, textiles and ironware from Britain and Northern Europe – all and more passed from the holds of inbound slave ships to African markets. There they were exchanged for enslaved Africans. Those ships also transported commodities produced by colonial economies in the Americas. It is no small irony that goods cultivated by African slaves in the Americas should find favour with African middlemen on the Slave Coast, and thence into the interior economies. Tobacco from Bahia (some of it soaked in molasses), and from Barbados and Virginia found a ready market among slave traders. So, too, did rum.

Local alcohol had long been important in any number of African societies; in religious ceremonies and social customs, as well as simply for pleasure. Imported alcohol, especially French wines and brandies, had also become favourites in West Africa, and were often handed over to influential Africans as gifts to facilitate trade and as part of a barter system. In 1680, a merchant from Barbados discovered that rum sold better than brandy on parts of the African coast, and the pattern was set. Now, prodigious volumes of rum were being shipped across the South Atlantic by Brazilian merchants, and by New England slave traders to West Africa – an estimated 300,000 gallons in each case. By the early eighteenth century, about one in seven British slave voyages started out from the Americas, and they carried rum, rather than the variety of goods shipped by traders leaving from British ports.

It is true that the popularity of rum on the African coast varied from place to place. In some locations, it was in great demand: parts of the Gold Coast imported 48,000 gallons a year *c.* 1700.[19] Along with other imported alcohols, rum entered African societies which had their own existing alcoholic drinks.

But these imported drinks had (like tobacco) an astonishingly ironic twist. It was, after all, a commodity produced in the Americas by Africans who had been shipped across the Atlantic in the opposite direction. Slaves cultivated the very item now being traded in Africa, for yet more slaves.

The people who actually made the rum – the enslaved Africans – naturally drank rum as part of their regular diet. Rum was doled out by planters as part of the slaves' regular allocation, with extra rations given out as a reward, and on high days and holidays. On Worthy Park Estate in Jamaica, slaves received upwards of three gallons of rum each year in the last fifty years of slavery.[20] Rum eased their burdens, rendered their breaks and free time more tolerable, and fortified them against life's miseries. In their turn, slaves so liked the drink that they also bartered for it, exchanging goods produced on their plots and gardens, or for items they had made in their free time. Some received rum in return for sexual favours. In towns throughout the slave societies, rum shops (often run by freed slaves, by people of colour and, most notably, by women) were common. Rum was a prominent item for sale over the counter by slaves with an entrepreneurial bent.

On both sides of the Atlantic, huge numbers of people of all ranks were rum-drinkers by the mid-eighteenth century. It was fundamental to the work of sugar plantations, its revenue vital to planters and to the colonial state alike. Rum constituted a major element of trade across the Atlantic – in all directions – and had established a distinctive niche for itself in a wide range of communities, from the slave quarters of the Americas to the slave markets of West Africa. It was vital to the way the Royal Navy conducted its business at sea, and to the lives of soldiers everywhere. Grog shops dispensed rum in abundance – from

Boston to Sydney. Rum was consumed, often in astonishing volumes, by native peoples in Brazil, the Caribbean islands and in North America. It had, in effect, become a universal drink of common people throughout the Atlantic world. And this had come about from a by-product of sugar production. The pleasures afforded by sugar were mirrored by the relaxation and resolve derived from rum. It strengthened, emboldened – and sometimes corrupted – its legions of consumers on both sides of the Atlantic.

9

Sugar Goes Global

WHEN THE BRITISH ended slavery in 1833, the patterns of sugar production on plantations and the culture of sugar consumption in food and drink were well-established features of the Western world. Wherever Europeans and Americans travelled, settled and put down local roots, they took with them the eating and drinking habits forged elsewhere, often adapting their diets to local conditions. Their attachment to sweetness in all things proved to be a persistent – and even a necessary – feature of life. Americans and Europeans needed their sweet coffee, while the British and their emigrant offspring needed their sweet tea or coffee wherever they lived, be it in on the American Great Plains, in Cape Town, Calcutta or Melbourne. These same people also insisted on sugary sweetness in their cooking and baking across the vast expanses of the Americas, in colonial outposts in India, Asia and Africa, or in the new settler societies of Australasia.

The century which followed the end of slavery in the Americas (which took place between 1833 and 1888) was, of course, characterised by new waves of imperial expansion of Europeans in Asia and Africa, of Americans in the Americas and the Pacific, and everywhere the essential military might of Western nations was sustained by the drinks and foods of their homelands. No army – or navy – could be sustained in its distant posting without heaps of sugar in the familiar drink and foodstuffs of their homelands. Armies, as Napoleon famously noted, march on their bellies; they also liked those bellies to be sweetened. Sugar was a basic ingredient for every US Army soldier from the time of the Revolutionary War to the present day. When the Continental Congress ordered food for the Continental Army, it included molasses and sugar. (Rum was added after the war, in 1790.) In the American Civil War, sugar was among the rations which sustained the Union Army.

And so it continued, in different forms, down to the present day. Candies – made largely of sugar – were supplied to American troops on the Western Front, and all the combat rations (C, D and K-rations) of subsequent wars – the Second World War, Korea, Vietnam and the Gulf War – included sugar as part of the essential foodstuffs to maintain men in the field.[1] To this day, British Army rations include plentiful supplies of sugar.[2] The love of sugar, and of sweetness generally, simply spread round the world in the wake of migrating humanity, and their armies.

The rising demand for sugar was driven principally by the massive increase in the global population. It almost doubled (from 978 million to 1.65 billion) between 1800 and 1900. There were ever more millions of mouths to feed, and tens of millions of people, now weaned on sweetened foods and drinks, required sugar (and other sweeteners) as part of their daily diet. In Europe, the home to sugar's modern popularity, the population increased by 76 million in the first half of the nineteenth century. By 1900, it had doubled in the space of a century.

The growth of the North American population was even more spectacular – in 1800, it stood at more than 5 million; by 1900, it was 76 million.[3] Sustaining this rising population was, of course, one of the major tasks facing governments everywhere. How to feed an expanding nation became a pressing political as well as a commercial and agricultural issue. Hunger and starvation were to become, in the twentieth century, important weapons in military conflicts. At critical junctures in both the First and Second World Wars, the combatants' ability to feed their populations – or starve their opponents into submission – became critical military tactics. Food was, obviously, a matter of life and death, and sugar was integral to the entire issue.

To provide the world's growing population with its necessary sugar required new systems and new areas of sugar cultivation. We shall see how the lands of the American Midwest were converted to beet production, but that alone was insufficient, even for the needs of the USA. Sugar cultivation and production spilled out into regions of the globe previously untouched by the commercial sugar industry.

Sugar cane could be cultivated in a number of tropical locations but, in 1800, its commercial concentration was effectively restricted to that scattering of Caribbean islands and Brazil. A century later, however, sugar was cultivated commercially on a huge scale, in all corners of the tropical and semi-tropical world. By 2000, more than 100 countries cultivated sugar cane.[4] The sugar plantation had been the pioneer of a new form of agriculture but, in the course of the nineteenth century, the plantation was transformed by the coming of modern mechanisation. The development of steam power and the establishment of central, highly mechanised sugar factories created not only a more efficient means of processing sugar cane, but those factories speeded up the whole process of sugar production. These new factories devoured increasing volumes of cane and therefore dictated work patterns in the cane fields. In order to keep the factories at peak capacity and performance, the acreages of land under cane increased substantially, and sugar plantations became bigger. In the mid-eighteenth century, a substantial sugar plantation might boast some 2,000 acres. By 1900, 10,000 acres represented a large plantation.[5] Much depended on topography, of course. Not every sugar-growing region was ideally suited for such massive operations, but where suitable land was available in large expanses, it was possible to develop very large sugar plantations.

Sugar thus crept into a great variety of tropical locations. Indeed, it was assumed by settlers and pioneers in many newly settled regions that the sugar plantation offered the best prospect for profitable agricultural development. The world wanted sugar, and history seemed to show that sugar was best cultivated on plantations. Once again, sugar became the key venture as new tropical lands were turned to commercial cultivation. There remained, however, one major problem, which had plagued sugar planters from the early days of Brazilian settlement in the sixteenth century – the thorny question of labour.

For all the mechanisation of sugar in the nineteenth century, sugar remained a labour-intensive crop. Before the very recent advances in mechanised cane-cutting (and even that was only suitable on large expanses of relatively flat land), the cane fields remained the preserve of back-breaking manual labour – gangs of men and women, bent double, toiling at the onerous work of planting, weeding, fertilizing and finally cutting the sugar cane. Whoever undertook that work – slaves or the freed labour that followed – it was physically extremely demanding.

It was also ill-rewarded, both by planters and by the corporations that came to dominate many of the sugar industries. Sugar-cane workers toiled hard for very little. It was no surprise when emancipation came for slaves in the sugar colonies that large numbers simply quit the plantations, ideally heading for their own small plots of land to work as free peasants, where they often produced sugar cane for the nearest sugar factory. Work in the sugar fields remained as difficult and punitive as ever, and labourers generally stayed on their old plantations only when there were few alternatives. As long as slavery survived (in the Caribbean, the USA and Brazil), sugar planters could find ways of augmenting the workers in the fields by buying more slaves.

Freedom created problems of a different kind for sugar planters. When planters, colonial officials and imperial governments turned their back on slavery (and often boasted about their virtue in doing so), they turned to a new form of imported manpower – indentured labour – which was itself much less than free. The new source of that labour was not Africa, but India. These labourers from India were shipped across the Indian Ocean and the Atlantic to plug the labour-force gaps in the Caribbean colonies. Again, the numbers were astonishing. In the ninety years to 1924, European powers, led by the British, had shipped almost 1.5 million Indians as indentured labourers to the old sugar colonies and to some new settlements. More than 250,000 went to British Guiana, and almost 150,000 to Trinidad. Large numbers were also transported to work on new sugar schemes in other parts of the world.[6] After 1879, for example, Indians were shipped to Fiji to work the new sugar plantations which had been established by Australians. A century later, by which time sugar dominated the Fijian economy, the descendants of those labourers formed half of the local population. A new ethnic hierarchy had emerged – but so, too, had deep-seated enmity between the indigenous peoples and the descendants of the indentured workers. Sugar cane, once again, scattered the spores of animosity far and wide; it created social and ethnic divisions and hostility from the Caribbean to the remotest of the Pacific Ocean islands.

A similar pattern unfolded in the Indian Ocean. In Mauritius (which was British after 1810), 455,187 Indians replaced former slaves in the cane fields. There, sugar planters, like their contemporaries in the Americas, increased their output by constructing large centralised factories, but their Indian labourers disliked the plantations and tended to move away when their indentures expired. Planters persisted, and continued to turn to India for still more labourers and, by the mid-twentieth century, their descendants had become a majority of the local population. Much the same happened in British Guiana and, to a less marked degree, in Trinidad. In all those places, Indian labour stepped into the jobs vacated by freed slaves on existing sugar plantations.

It was different in South Africa, where local sugar cane had long been popular among Africans. But the British settlers introduced new, commercial varieties of cane, which they cultivated on plantations in Natal. Africans, however, proved reluctant to adapt to the rigours of plantation labour and the British, after 1860 (and following the Caribbean example), once again turned to Indian indentured workers. Natal's sugar exports boomed, though ill-treatment and harsh working conditions remained a perennial complaint among the Indian labourers. Even so, South African sugar thrived, greatly helped by the extension of British political power across stretches of southern Africa, and by favourable terms (in land and taxes) granted to sugar planters. After the annexation of Zululand in 1897, the authorities controlling local land granted vast tracts to sugar planters. Working with local sugar millers, and sustained by British policy, the South African sugar industry became a major producer not only for South Africa itself, but for the wider British market.

Through all this, the sugar industry continued in its own distinctive, destructive ways. Even when indentured Indians quit the South African cane fields, they were replaced by different migrant workers, this time from other parts of South Africa. They, too, endured a miserable fate, made worse not only by the harshness of the cane-field labour, but by the neglect and brutality of their employers. This whole process was part of a broader South African saga, of the stark segregation of peoples – Indian, African and white – a segregation encouraged both by sugar interests and by the imperial government. Sugar, once again, proved its corrupting potential, scattering the seeds of ethnic turmoil that was to plague South Africa from that day to this. The young Gandhi, moving to South Africa, was radicalised by precisely this toxic cultural and economic mix.

Sugar was also developed in Australia's tropical north. Queensland's climate was ideal, and a local sugar industry would satisfy the very sweet tooth of Australia's expanding immigrant population, needing no longer to involve importing sugar vast distances at great expense. From the late 1860s, Australian sugar expanded around the northern town of MacKay, and it, too, adopted the plantation system.

By the 1880s, after a string of setbacks, Australia's sugar industry thrived and it, too, turned to migrant labour: Chinese, Japanese, Javanese and, above all, people from the Melanesian islands, mainly Vanuatu and the Solomon Islands. Some were indentured (on the Indian model), but that later fell away in the teeth of local white opposition. These migrations – known as 'blackbirding' – ranged from outright violent seizure of the labourers, to indenture and, once again, it created a form of labour for the cane fields that was less than free. It was only ended in 1904–06, but it had laid the early foundations of what became, a century later, a massive, highly mechanised Australian sugar industry. In the 1890s, Australia produced 68,924 tons of sugar. A century later, it disgorged 5.25 million tons, most of it for export. By then, Australians had established themselves among the world's greatest consumers of sugar – they consumed 48.34kg each per year.[7]

Wherever sugar plantations took root, they brought about massive changes in their wake. They changed the local ecology, altered local demography and transformed the politics of society at large. It is a story which was repeated time and again from one sugar economy to another, across the centuries, and from one corner of the globe to another – from São Tomé to Brazil, from the Caribbean to the Indian Ocean, from the Pacific to Australia. Long before it was fully appreciated how sugar helped to corrupt the physical well-being of consumers, it had exercised an astonishingly corrupting influence on the environment and populations of swathes of the inhabited world. And this was in addition to the volatile, political brew generated by sugar.

By c.1900, sugar plantations had proliferated around the globe. Once, they were to be found overwhelmingly in the Americas but, by 1900, they thrived worldwide, from Mauritius to Fiji, from Hawaii to Australia. Sugar cultivation had become a truly global industry and sugar plantations had, yet again, proved their ability to tap the commercial bounty of the world's tropical and semi-tropical regions. It was a process which continued at an even faster pace in the course of the twentieth century. Sugar was now consumed by societies the world over – and sugar was cultivated and produced wherever land and manpower could be combined to bring forth sweetness.

10

The Sweetening of America

BETWEEN THE MID-NINETEENTH century and the First World War, the Western world experienced seismic changes in its food and drink. What happened in the USA, however, was to prove especially important for the world at large in later years. The rise of the US economy and the global spread of US corporate power began to exercise an unprecedented influence over economic and social life in the world at large. The changes in America's diet, for instance, were later adopted around the world and, at the centre of the American diet, lay the story of sugar. Even by 1914, Americans (and Europeans) were devouring unprecedented volumes of sugar. By the mid-twentieth century, those sugary habits had taken root and flourished to an astonishing degree all over the world.

The most obvious reason for the increased consumption of sugar, led by Europe and North America, was the increase in population. Between 1800 and 1900, the world's population almost doubled; and as we've seen previously, North America's population increased from 7 million in 1800 to 82 million in 1900, reaching more than 300 million in 2000. In the same period, the people of Europe increased from 203 million to 408 million, and then to 729 million. Although major regional and national differences lay beneath these numbers, one simple point stands out – many more millions of people required food and drink. The world's resources were tapped – and taxed – to find sufficient sustenance for human needs and pleasures. In the process, food and drink were cultivated and manufactured on a scale never before imagined. New industries and technologies were given the task of devising new methods of food cultivation and food processing to satisfy an ever-growing demand. As with almost every other walk of life, new industrialised processes emerged to face the challenge.

In agriculture – be it arable or livestock – and fishery, revolutionary large-scale methods and equipment began to change the way crops, animals and fish were cultivated and reared – then processed. Transforming those products into marketable foods demanded new industrial systems of processing, packaging, distributing and marketing. Some of those changes are familiar. Railways, for example, had a dramatic impact, reaching into distant farmlands of North and South America and Australia, and bringing tons of grain and livestock to the grain elevators and slaughterhouses of major port cities and, from there, they were exported to Europe and beyond.

Behind this story of modernised food and drink lay a major fall in the price of key commodities. People could buy more with their money. The American dollar could buy more food, say, in 1900 than in 1870. Sugar prices, which had been falling since the Civil War, fell even further, thanks largely to industrialisation of sugar refining.[1] This story, of the cheapening of food as it became more industrialised, is very familiar because it is the essential heart of modern foodstuffs. Less clear, in what was to become a long, drawn-out process, is why sugar should remain central to the entire story. Why and how did sugar (and, later, other sweeteners) intrude themselves into the new drinks and foodstuffs of the West, and then of the wider world?

Even in its early stages, the industrialisation of food and drink in the late nineteenth century involved the widespread use of sugar. Sugar was used in the development of new types of flour, in the packaging and refrigeration of meats, and especially in the new canning and bottling systems of a wide range of fruits, vegetables and soups. Step by step, sugar found its way into an enormous range of foodstuffs delivered to people's dining and kitchen tables. In the same period, the rise of cheap fizzy drinks – sodas – had an even more prominent role to play in the use of sugar.

To understand how this happened, we need to look at the USA. Throughout the nineteenth century, the US economy, and the people it served, consumed sugar on a gargantuan scale. That sugar was imported from the Caribbean and later from Hawaii, while millions of tons were cultivated in the cane fields and sugar beet fields of the USA. On the eve of the Civil War, the US imported 694 million pounds of sugar. Twenty years later, that had risen to 1,830 million pounds. In 1900, sugar imports stood at 4,007 million pounds, only to double again by the mid-1920s.[2] While the US was not alone in its love of sugar, its economy was hugely influential in shaping and influencing others around the world, and the story of what happened to sugar in the USA foreshadowed many of the global trends of the twentieth century. Today, key areas of world food and drink are dominated by major US corporations and, if we want to understand the global patterns of sugar consumption, we need to look closer at what happened in the USA.

Earlier, in Chapter 6, we examined the story of coffee-drinking in the USA, and the American love of sweetened coffee that became, and has remained, a national obsession. But sugar itself had a far wider importance in the USA than simply being an additive to coffee. Sugar became an issue in US federal economic policy, with tariffs imposed on imported refined sugar in order to protect the US sugar refining industry – and, of course, to generate income for the Government. Although the first tariffs were in place in the early days of the Republic, protection for the sugar industry only came fully into force in 1842. Fifty years later, when the McKinley Act (1890) changed the tariffs on sugar, sugar prices fell, and sugar consumption increased even further.

Until the 1880s, US sugar had been characterised by a multitude of sugar companies, with their refineries dotting the port cities of the east coast. The modernisation of the industry in the 1880s was prompted by major investments in new refining facilities and by the emergence of major refining companies led by the flamboyant entrepreneur Henry Havemeyer. He had been born into a family sugar business, was trained and raised in sugar, and was to prove both an astonishing businessman and the driving force behind the transformation of the American sugar industry. In 1887, Havemeyer amalgamated seventeen of America's twenty-three refineries to create the American Sugar Refining Company – which controlled 98 per cent of the industry. It took the form of a trust – the Sugar Trust – mirroring Rockefeller's more famous Standard Oil Trust. Like other trusts, Havemayer's version revolutionised the sugar industry and, by 1891, there were only four of the original refineries in operation.

These were the years of the formation of major corporate conglomerates – the rise of Americans trusts – in a string of industries: banking, tobacco, finance, oil and steel. These trusts were so massive and powerful that they effectively operated a stranglehold – in some cases, a monopoly – over their industry. But they prompted a fierce political and legal reaction, a conflict between the major industrial trusts (keen to press home their domination and to resist the legal and political restraints placed on them) and politicians anxious to defend the consumers' interests. And it was a long-running story, a part of the 'momentous organisational convulsion' that transformed the American economy between 1895 and 1907. It was an era of major takeovers – the yearly average was 266 – and the end result was normally a single company dominating its chosen market. Sometimes, a single corporation might control 60 per cent of the market. By 1904, an estimated 318 corporations owned 40 per cent of all US manufacturing assets.

The Sugar Trust, formed in 1887, only five years after Rockefeller's initiative in oil, was reformed in 1891 as the American Sugar Refining Company. It consolidated the major sugar refiners of the USA, and became the central agent for the manufacture of American sugar, owning most of the nation's best-known brands, and was responsible for negotiations with state and regional governments. Havemayer's company, with various subsequent legal names, became the main organisation for the manufacture of American sugar, and concentrated on promoting highly refined, white, granulated sugar. It also did no harm, from Havemayer's point of view, to set out to undermine other sorts of sugar, notably brown sugar, in the process.[3]

Throughout, it developed complex but persuasive dealings with government.

As with other major trusts, it, too, had serious conflicts in the law courts and with government. But its power, and its money, enabled this new brand of American 'King Sugar' to deflect or win over political and legal opposition and to advance the broader interests of the sugar industry. Although sugar is rarely lumped together with railroads, steel and petroleum, its cartel arrangements followed a very similar path – and with similar consequences for the way the industry was run. It was also a vivid illustration of the importance of sugar in the economic and social life of the USA at the dawn of the twentieth century.

This corporate story of American sugar, and its growth in the last years of the nineteenth century into a major national industry, was also a striking reflection of America's deep attachment to sweetness. The sugar lobby and industry were important because sugar was an integral feature of American life. Sugar was everywhere – from medicine to home cooking, from commercial drinks to elaborate dining at family and formal occasions. Many of these sugary tastes emerged as powerful commercial entities from older, simpler and more functional items. The sweetening of bitter medicines, for example, led to the emergence of sugary candies (sweets). American candy bars were little more than boiled sugar.

By the early twentieth century, sugar, a long-time presence inside the home as an additive to drinks, became a staple in kitchens and cupboards as a vital ingredient in home cooking, baking and preserving fruits and foodstuffs. Sugar was equally important in the massive expansion of commercial baking and confectionery.

A new breed of commercial entrepreneurs, men who spotted the demand for sweet, luxury tastes, developed the highly mechanised production of modern sweets and chocolates, and were inspired to sell them in small, individual packets or items – a lollipop, a chocolate bar. Milton Hershey, following European pioneers, launched what became the massive American chocolate company that bears his name. So, too, Frank C. Mars, whose chocolate company became one of America's largest family businesses. Candies and chocolates in myriad shapes and sizes were blended from a multitude of ingredients – but always with sugar. They, too, became essential components of American military rations from the First World War onwards. Sweetness offered instant enjoyment, strength and resolve – and, perhaps, brought fond reminders of very distant home pleasures and comforts for men serving overseas.

As the volumes of home-grown and imported sugar increased, the price of sugar in the USA dropped, and sugar-based rituals became an inescapable feature of the American diet. Sugar was essential, even for pioneers heading west. Printed guides to the perilous trek across country in the 1840s, for example, listed all the foodstuffs required for the journey – it included a supply of salt, coffee and sugar.[4] When Mark Twain travelled by stagecoach from Missouri to Nevada in 1861, he complained that, at one stagecoach stop, he was served a foul concoction instead of tea – and no milk or sugar to go with it.[5] Cowboys herding cattle on the plains in 1881 made sure that their food wagon contained plenty of sugar alongside other necessities. In all corners of the USA, from the most fashionable of New York City salons to life on the settler and cowboy frontiers, sugar was essential.[6] Americans needed sugar in their breaks at work, at rest between chores and when travelling across that mighty landscape.

One of the most distinctive American inventions based on sugar that has passed down to us today is the elaborate, iced and tiered wedding cake. Today, we take the wedding cake for granted, but it emerged as a popular sugary indulgence in the last decades of the nineteenth century. The icing and decorations were almost exclusively sugar-based. It was, after a fashion, a modern-day version of the sugar sculptures of earlier periods, although, at the time, such displays were signs of exclusivity. Now, the sugary, white wedding cake had been democratised. Of course, many older communities had celebrated weddings with cakes with exotic ingredients, but the modern habit took off in its contemporary, sugar-laden form in the USA. By the 1830s, American recipes showed how to dress a wedding cake in almond paste, made with sugar, and, as the century advanced, these cakes became increasingly elaborate. The wealthier the families, the more ornate the cake, some of them baked by immigrant German bakers.

By the end of the century, cheaper ingredients, especially flour and sugar, commercial mass-production, new ovens, advertising – and rising spending power – enabled the American middle classes to enjoy wedding cakes of a kind they could once only have dreamed about. Now the cakes were coated in white sugar icing made from the most refined sugar, and became more elaborate – they were finally stacked in tiers. Queen Victoria's wedding cake was 9ft in circumference, but only the base was cake – the upper tiers were made entirely of sugar. The cake would be cut by the bride and groom, a ceremony started in the mid-nineteenth century, but which had become universal by the early twentieth century. All this was made possible by the abundance of cheap sugar.[7]

The wedding cake was only the more elaborate form of a more general pattern. At much the same time, sugar became a key ingredient in a huge number of desserts which began to dot fashionable tables and menus in late-nineteenth-century America – gelatins and jellies, as well as ice creams, bought commercially or, thanks to the spread of refrigeration, made in more prosperous homes. Sugar also featured as a main item on America's evolving calendar of holidays and festivals – sweet treats at Christmas and Easter, on birthdays and St Valentine's Day events. All were celebrated by sugary foodstuffs, and all were to reach remarkably elaborate heights in the late twentieth century.

In this, as in all dietary matters, American food was transformed by the introduction of industrialised, packaged foodstuffs and refrigeration. How to keep food, to stockpile or preserve it, had been a problem from the earliest days of settlement. European immigrants were unwilling to follow the Native Americans in accepting that there were periods of the year when hunger prevailed until the seasons changed. Preserving foods for winter required a plentiful supply of vinegar – and sugar. Pickled vegetables stood alongside sweetened fruits and jams in homes across America. It was a golden rule of preserving fruit to use plenty of sugar. One suggestion was that the best way to preserve jam, jellies and conserves was to use at least 60 per cent sugar.[8] One cookbook claimed, in 1844, 'If too small a portion of sugar is allowed to the fruit, it will *certainly* not keep well.'[9] As long as sugar remained costly, it was more economical to dry fruit rather than preserve it with sugar but, when sugar became cheaper, and available in great volumes, it helped to sustain North Americans through their harsh winters. This, too, changed after 1858 with the invention of the Mason jar, a glass jar that was sealed via a screw-on lid and rubber seal. By keeping out fresh air, the jars required less sugar for preserving fruits, but even this innovation was soon replaced by the development of commercial food canning.

The initial process of food canning began in France, and was rewarded by Napoleon who recognized its potential for feeding armies on the move. By the 1820s, it had spread to Boston and New York, where the commercial process, at first, concentrated on bottling fruits and ketchup, before canning became the main process. From the first, sugar was added and followed traditional recipes for making ketchup in the home. Now, however, it was produced by the tens of thousands of gallons.[10]

At first, the process of canning was laborious because it was largely undertaken by hand, and the final product was costly. It also came with early health risks, such as contaminated food and poisoning. All this was swept aside by the Civil War, and the Union Army's massive purchases of canned foodstuffs for the troops. Numerous entrepreneurs rose to the challenge and set about improving the system, and making the contents safer to eat. At the end of the war, when the Army released hundreds of thousands of men back to civilian life, those men had become accustomed to, and generally fond of, the range of canned foods they had eaten as soldiers. For many of them, particularly poorer men, the army diet had been more varied and more pleasant than the meals they had consumed before the war.

Along with everything else, the Civil War transformed the canning industry. In 1860, the US food industry produced 5 million cans of food. A decade later, it produced 30 million. As the canning business became increasingly industrialised, its products brought an expanding range of foodstuffs within the reach of ordinary Americans. No less important, the new railway system enabled foods to criss-cross the length and breadth of that vast continent. In the process, regional foods became national. Meat and flour, fruit and vegetables were packaged cheaply, and could be bought cheaply thousands of miles from their point of cultivation or production. And the same was true of sugar.

This post-1860 packaging of foodstuffs wooed consumers by its simple convenience. What today seems natural was, in the late nineteenth century, a change of fundamental proportions. The Campbell Soup company, for example, which started in 1869 with canned vegetables, had, by the turn of the century, concentrated on soup production. By 1904, it was producing more than 16 million cans a year. All was accompanied by extensive advertising of the company's products, and a parallel campaign to persuade customers that their foodstuffs were healthy. These were the great, pioneering days of modern advertising – led by the US tobacco industry – and for all the colour and extravagance, the visual and verbal excesses of such campaigns, many of them concentrated on the healthy nature of canned products. From Campbell's soups to Duke's American Tobacco Company, the consumer was lured not only by the convenience and cost, but by the claims that the products would aid well-being. However unsubstantiated or deceptive the claims of this advertising, American customers bought the new foodstuffs in huge and growing volumes.

In the case of these new canned foods, whatever loss of nutrition may have been incurred in the technical processing – along with the addition of sugar – was more than compensated by the ease of purchasing, preparing and consumption. It also made the kitchen a less laborious place, promoting the short cuts to cooking via a welter of pamphlets and cookbooks which concentrated not only on a company's own products, but on the most convenient methods available to cook and feed a family.[11]

What had emerged, by the early years of the twentieth century, was a totally new culinary and commercial phenomenon – the cult of cooking convenience. Within the space of fifty years, the diet and the cooking of the world's most important and dynamic economy – the USA – had been successfully won over to the ideal of convenience. Food and drink was available cheaply, in a variety of convenient formats – in bottles, cans and packages. They were within the reach of everyone and, above all, they offered the irresistible promise of easy preparation to the womenfolk who were the traditional providers of meals everywhere.

More than that, the stage was set for a much more thorough and radical industrialisation of food production and marketing throughout the next stage of the twentieth century. In the process of creating these early convenience foods, sugar had been the vital helpmate. In the twentieth century, sugar – and, later, artificial sweeteners – were to change not only American habits, but the diets of millions of people the world over.

II

Power Shifts in the New World

FROM THE SEVENTEENTH century onwards, sugar was a source of controversy in domestic politics, a contentious issue in fiscal matters and even the cause of strategic, diplomatic and military conflicts. Sugar taxes were a bone of contention in the eighteenth century and, above all, a source of conflict between Britain and her American colonies. For the best part of two centuries, sugar-producing islands were also the source of conflict between feuding colonial powers. Indeed, Europeans shaped their foreign policies around their determination to acquire sugar colonies – or to prevent their rivals from doing so. In the process, major conflicts were launched, sometimes with disastrous effects, and global power tilted one way and then the other. The European struggle for sugar-based power in the Caribbean reached its extraordinary climax in the years of the French Revolution.

The value of the Caribbean to the Europeans was enormous. On the eve of the French Revolution, Britain's islands exported produce worth £4.5 million to Britain, and this at a time when Britain's own exports totalled £14 million. The French case was even more striking. France exported £11.5 million of goods, while their Caribbean colonies exported goods to the value of £8.25 million. This involved, of course, a range of tropical goods (cotton and coffee, for example), but sugar dominated in bulk and value.

France's Caribbean colonies were supremely important by the late eighteenth century. St Domingue (Haiti) had become the world largest sugar producer by 1770. And although the French loved their sugar, they re-exported 70 per cent of the Caribbean sugar, mainly to Holland and Germany. To this day, the impact of France's sugar economy can still be seen in the city that became the beating heart of France's Caribbean trade – Bordeaux – with its lavish buildings and elegant streets, and its industrial and commercial activity stretching along the Garonne and deep into the wine-producing hinterland. A similar story can be told of Nantes, the centre of the French slave trade, and Le Havre. Even in distant Marseille, one fifth of that city's trade was with the Caribbean. In 1791, Rochefoucauld calculated that more than 700,000 French families owed their livelihoods to sugar. Unlike the British trade, however, the French sugar trade was fragmented and, like the sugar islands themselves, individual French ports viewed other sugar ports as rivals. And *all* of them viewed the French state – with its taxes and restrictions – as an obstacle to be overcome, not as a co-operative partner. Much the same was also true of French sugar refineries; they, too, were pitched one against another. The end result was that the French sugar industry did not develop, as the British did, into a powerful and influential lobby able to influence government and policy. It was a fiercely divided business.[1]

For all that, the French domestic demand for sugar was widespread and growing, with sugar widely used as a basic ingredient in French cooking, in beverages, as a preservative, in brewing, in medicine and alcohol. The centre of French sugar consumption was in and around Paris, and Parisians consumed anywhere between 30lb and 50lb of sugar each year. More refined sugars were favoured by the rich, cruder sugars by the poor.[2] On the eve of the French Revolution, in both France and in Britain, commentators were united in feeling that there was an untapped demand for yet more sugar. The market seemed boundless. With sugar firmly at the heart of both British and French life and politics, any conflict between the two nations in the Caribbean had complex ramifications. Threats to each nation's sugar trade, disruption of shipping – of goods and African slaves – attacks on opponent's sugar colonies or, worst of all, from everyone's viewpoint, revolt by the slaves, was much more than a colonial issue. Interruption of the sugar supply meant serious damage to colonial power, domestic prosperity and well-being, and neither France not Britain could afford, nor even contemplate, disruption to their sugar economies.

French colonial power was thrown into turmoil by the Revolution in 1789 and by the subsequent slave uprisings in the French islands. A trail of destruction swept through the French Caribbean, but what happened in St Domingue was the stuff of nightmares for slaveholders everywhere. Insurgent slaves destroyed hundreds of sugar and coffee plantations and 10,000 slave-holders and refugees fled, mainly to Jamaica and the USA. The end result was the devastation of the French sugar industry. In 1791, St Domingue, at that time the world's largest producer of coffee, had also produced 80,000 tons of sugar. Within a decade, all that had collapsed, with sugar down to 10,000 tons, while Jamaica's had risen to 100,000 tons. France's booming sugar industry and its major source of tropical trade had disintegrated.[3]

France now had to look elsewhere for its sugar supplies, or seek alternative sources of sweetness. Eventually, Napoleon ordered the promotion of beet sugar production, although this was not as revolutionary as it sounds. Cooks had long known that the boiled juice of beet was similar to the syrup from sugar and, in the early seventeenth century, scientists had confirmed that boiled beet produced a sweet syrup.

But it was a German scientist, Andreas Sigismund Marggraf, who managed to produce sugar crystals from beet in 1747. One of his students, Karl Franz Achard, financed by King Wilhelm III of Prussia, took the process much further, showing how beet could be converted to sugar in large volumes, as well as being produced commercially. Achard's work continued after his death at experimental factories in Silesia and then Paris, but the turning point was France's Caribbean disaster and Napoleon's anxiety to find a new source of sugar.

Sugar beet was an innovation with enormous consequences. The French hoped (and the British worried) that beet sugar would undermine the British sugar industry. European consumers saw it as an alternative to their dependence on imported sugar from other nations' tropical colonies.[4]

Napoleon lavished praise on the initial product, predicted the end of Britain's sugar dominance, and ordered 100 students to be sent to newly established sugar-beet schools. He compelled farmers to turn to beet cultivation and made available 80,000 acres of land, experimental facilities and 1 million francs for the development of sugar beet by French farmers and manufacturers. The outcome was remarkable. By 1812, forty French factories converted 98,813 tons of beet into 3.3 million pounds of sugar. The cultivation of beet and the manufacture of beet sugar was quickly taken up by other European nations, notably in Germany.

The end of the Napoleonic wars in 1815, and the re-opening of European ports to cheap sugar imports from the colonies, saw the collapse of the fledgling beet industry. Though beet sugar simply could not compete commercially with slave-grown cane sugar, a breakthrough had been made; it now seemed possible that the world's sweet tooth *might* be satisfied without recourse to cane sugar shipped vast distances from the tropics with all the logistical and political difficulties involved. A new science and industry emerged devoted to sugar beet, all of it located in temperate countries. Scientific innovations led to major improvements in the processing of beet, but as long as cane sugar was cultivated by slave labour (which was not ended by the French until 1848), cane sugar retained its commercial edge. In addition, the colonial sugar producers held consider-able political power in Europe's capitals and were usually able to secure legislation to support their cause.

Germany, on the other hand, had no sugar-producing colonies, and enjoyed a steady development of local beet sugar. In 1836, there were 122 sugar-beet factories in Germany, and consumption rose steadily throughout the nineteenth century. By 1886, the German industry produced 1 million tons of sugar, more than doubling again by 1906. By then, the sugar-beet industry had taken off across Western Europe, from Belgium to Russia, with many hundreds of factories producing beet sugar for local consumption.[5]

The development of sugar from beet was initially a European project with countries across the continent striving to cater for the sweet tooth of their expanding populations. The USA followed suit. Sweet food and drink were vital factors in the diet of America's expanding population. Huge volumes of cane sugar continued to flow north from the Caribbean, notably from Cuba and the other Spanish islands and, later still, from Hawaii. But beet sugar offered the USA the tantalising prospect of home-grown sweetness. There were, moreover, enormous expanses of the USA which seemed ideally suited to beet cultivation.

There were early attempts to develop sugar beet in Pennsylvania and Massachusetts and, later, by Mormons in Utah (with equipment bought in Liverpool). Further trials followed in Illinois, Wisconsin and California, with factory machinery shifted from one location to another as the beet experiments moved around the country. Most failed to make money until, in the late 1880s, factories in California became profitable. By the end of the century, new enterprises in both Nebraska and California firmly established US sugar beet as a profitable and thriving business.

Encouraged by new taxes on imported sugars, American-grown beet sugar boomed. In 1892, there were six factories in the USA producing 13,000 tons of sugar. Ten years later, forty-one factories were disgorging more than 2 million tons. By the mid-twentieth century, the US sugar-beet industry was a highly mechanised affair, producing 3.5 million tons of sugar, around one quarter of America's needs. It was, by then, dominated by major corporations with sixty plants scattered across eighteen different states, and producing 100 varieties of refined sugar. In the early years of the twenty-first century, it received government support to the tune of an estimated $1.6 billion.[6] Yet for all the importance of sugar beet, the USA in the late nineteenth century continued to devour cane sugar from tropical growers, especially from the Spanish islands in the Caribbean.

Despite the science and modern processing techniques used in the production of beet sugar, the cultivating and harvesting of the beet itself involved miserable, physical toil. Not on a par with slavery in the sugar fields, it was nonetheless miserable work in harsh conditions. It was yet another illustration of the paradox at the heart of sugar – harsh labour was needed to produce a commodity which was essentially a luxury item. Until the seventeenth century, people had managed without sweetness in their food and drink, but now, thanks to the rise of sugar slavery in the Americas, the world had become dependent on sweetness in all things, whatever the cost to the labourers who endured hardship to produce the agricultural crops involved.

At every turn of this story – cane or beet – the cultivation and processing of sugar remained a hotly contested issue. Long after the demise of the old colonial powers and the rise to dominance of the USA, sugar remained a highly politicised commodity. In many respects, the USA followed in the footsteps of Europe's old colonial powers and, in time, came to exercise its own (even greater) power in pursuit of its strategic sugar policies in the Caribbean and, later, in the Pacific. However sweet to the tongue, sugar was responsible for a bitter political aftertaste. Americans, like Europeans before them, became wedded to sweet food and drink, and their politicians sought to defend and promote American sugar interests.

Sugar was playing an important role in US politics by the late nineteenth century – even in foreign and strategic affairs. Like Europe in the eighteenth century, sugar was deemed so important to American well-being, so vital to what Americans ate and drank, that any threat to US sugar interests became a weighty political issue. US foreign policy developed a special interest in sugar-cultivating regions (especially in the neighbouring Caribbean islands) and even in the Pacific. Hence, at critical junctures, key US foreign policy issues were shaped around the question of sugar.

In 1897, a major US journal commented that sugar had become 'the American question of the day'. The article stated, 'To sugar or not to sugar seems to be the present issue in the United States Senate.'[7] Questions of sugar cultivation, of sugar imports and refining, the problem of sugar supplies and prices – all these had, by the 1890s, forced the commodity to the top of the US political agenda. And there it was promoted by a powerful commercial lobby whose influence reached far beyond US borders, and whose political power was hard to ignore in Washington. But everything hinged on one simple fact – the American people consumed sugar in vast and increasing volumes. Unlike the old European powers, though, the USA had the capacity to grow its own sugar.

* * *

There had been attempts to cultivate sugar in colonial North America, in Virginia, in Georgia and South Carolina, but most of those efforts had been unsuccessful or uneconomic. The most suitable region for cane cultivation was Louisiana but, even there, experiments with imported Caribbean cane in the mid-eighteenth century proved a commercial failure. Louisiana sugar only began to flourish in the 1790s in the wake of the Haitian revolution, and then, more substantially, when Louisiana was acquired by the USA. Until 1803,

Louisiana had, by turns, been Spanish then French. The Louisiana Purchase from France in 1803 – for a mere $15 million – was to prove an astonishing bargain, heralded, even at the time, as a master stroke. President Jefferson was congratulated on the deal by General Horatio Gates, who said, 'Let the Land rejoice, for you have bought Louisiana for a Song.' The deal *doubled* the landmass of the USA and, among other benefits, opened the potential for enormous agricultural development in the fertile lands of the Mississippi Delta. Although the land seemed ideal for sugar cultivation, it badly needed labour for the arduous, intensive work in the sugar fields. But unlike sugar planters in the Caribbean and Brazil, the new American Republic had no need to turn to Africa for labour because there was a growing population of enslaved people scattered around the Old South. For a price, that labour could be transferred south and west to the new cotton and sugar industries along the Mississippi.

A new breed of overland US slave-traders began to move American slaves across state lines to new settlements, to develop Louisiana's potential. Unlike General Gates, the slaves had little to rejoice about. They were to endure the miseries of a lifetime in the sweltering heat of the new cane fields and sugar factories, and to bequeath their sufferings to their children, much as their forbears had done in Brazil and the Caribbean.

This time, however, sugar slavery was different. For a start, the USA was a rapidly modernising country, able to harness new industrial technologies to innovative and intrusive management systems. The Louisiana sugar industry quickly established itself as a unique mix of old and new – old-fashioned, brutalised slave field labour, kept at work by new machinery and modern management.

In 1812, Louisiana had a mere seventy-five sugar mills, but the introduction of new types of sugar cane – better suited to the climate and ecology of Louisiana – helped to bring about a rapid increase in cultivation. Louisiana's sugar estates also thrived thanks to the availability of capital, notably from banks both in New Orleans and in Europe. In the form of their slaves, the sugar planters were able to offer substantial collateral against their loans, and Louisiana sugar planters embarked on a major campaign of investment in modern sugar equipment. By the time of the Civil War, Louisiana's sugar planters had overseen the most heavily invested form of agriculture in the entire USA.[8]

Louisiana's sugar lands, like the neighbouring cotton plantations, depended on vital steam-driven river transport, and they also harnessed steam power to the cultivation and processing of sugar. Yet at the heart of this modern sugar industry was enslaved labour. The country's rapidly growing demand for sugar prompted a major expansion of sugar production. The number of sugar estates doubled in the late 1820s; production surged even more in the 1840s, when the older properties were joined by new sugar estates west of New Orleans. These were boom years for Louisiana sugar. One planter spoke of 'the sugar gold fields'. Within a mere sixty years of taking firm root in Louisiana, sugar boasted an estimated 1,536 sugar estates producing 250,000 hogsheads (about 125,000 tons) of sugar. At mid-century, they produced more than 320,000 hogsheads (about 160,000 tons) of sugar. A year later, Louisiana made one quarter of the world's sugar exports, although natural disasters hit the industry hard in the late 1850s. The last sugar crop to be harvested entirely by slaves – in 1861 – yielded 460,000 hogsheads (about 230,000 tons) of sugar.[9]

Throughout these boom years, Louisiana planters had been innovative, experimenting with different types of cane, trying new cultivation systems and modern ways of processing it. They were also helped by the falling price of new equipment which resulted from innovations in the metal and engineering industries in the northern states. The outcome, on the eve of the Civil War, was that the Louisiana sugar industry, like the Lancashire cotton industry, was driven forward by steam power and the latest technical innovations.

Yet all this went hand in hand with the expansion of slave labour. In 1827, some 27,000 slaves worked in Louisiana's sugar fields; by 1850, the figure stood at 125,000. In addition, the slave gangs employed on Louisiana's sugar estates grew significantly. The average slave gang in 1830 numbered fifty-two but, by the Civil War, it was upwards of 110.[10] On the smaller, less modernised sugar plantations employing slave gangs of twelve or so, the enslaved endured perhaps the worst conditions, sometimes living in buildings no better than ox sheds. Set alongside sugar estates in the Caribbean and Brazil, these were relatively small numbers, but compared to the typical slave-holdings elsewhere in the USA, they were large. Even so, the images of slaves at work in the cane fields of Louisiana would have been recognizable to sugar planters a century before – lines of bent backs slashing at a wall of tall, waving sugar cane, using that universal tool of sugar cultivation – the machete. Theirs was a daily working routine characterised by extreme hardship, and always under the threat of violence by overseers and managers. The irony, however, was that now, in the nineteenth century, the most modern machines available were kept at their steam-driven routines by field slaves. Cane which lay in the fields too long, or machinery with no cane to process, meant inefficiency and financial loss.

Machines in the sugar factory demanded a regular flow of sugar cane to keep the process in motion and sugar planters and their managers had to find new and ever-more persuasive means of keeping the field gangs hard at work. Planters and outside observers began to *admire* the efficiency of the Louisiana system. Few seemed to notice that the real price was paid by an enslaved labour force pushed to the extremes of physical endurance.

These nineteenth-century sugar estates were distinguished by their efficiency and their profitability. Modern machines worked hand in hand with a new form of slave management, and sugar planters came to pride themselves on the resourceful management of their sugar estates; some even viewed the plantation as a machine itself. It all served to yield lavish returns to successful planters and, like Caribbean sugar planters before them, US sugar men were keen to flaunt the fruits of their efforts by building the most elaborate homes. Their residences, at the heart of the plantations, adopted the extravagant styles of self-made men on both sides of the Atlantic. Sweeping driveways, Palladian entrances, the finest of domestic furnishings and fittings (an image perhaps best represented by Tara, the cotton plantation in the movie *Gone with the Wind*). All this sat precariously cheek by jowl with impoverished slave quarters and an enslaved labour force obliged to endure misery on an epic scale.

The benefits also fell to American consumers who added the sweet product of slave labour to the huge volumes of imported Brazilian coffee, itself cultivated by slaves. In what was an uncanny reprise of the European experience in the seventeenth and eighteenth centuries, slavery continued to shape the sweet habits of the Western world.

* * *

As with many other areas of American life, the Civil War laid waste to Louisiana's sugar industry. By 1864, the number of plantations had fallen from 1,200 to 231 and the peak production of 264,000 tons of sugar had collapsed to 6,000 tons. But it brought freedom for the slaves, though many were reduced to abject conditions. Like other slaves throughout the Americas, when freedom came in Louisiana, few wanted to return to their old employment. The industry (and its former workers) were on their knees. Within a decade, however, Louisiana sugar had begun a major recovery. Northern money moved in, and nine tenths of the sugar lands changed hands.[11] New systems of cultivation and manufacture, central sugar factories, producers' organizations, research schools – all and more totally transformed a moribund industry. This modernized US sugar industry sought to make the most of the expanding demand for sugar both in the US and abroad. By the last years of the century, Louisiana's sugar industry was thriving and produced 302,778 tons in 1900.[12]

The world's sugar economy had changed fundamentally by then. America's expanding population, eager for ever more sugar in its food and drink, had a host of worldwide sugar producers, all keen to satisfy the American sweet tooth. Most importantly, some of them were very close to hand, in what was increasingly regarded as Uncle Sam's backyard, in the Caribbean. This combination – of American hunger for sugar and the proximity of sugar producers in the Caribbean – was largely responsible for establishing the US political and strategic interest in the Caribbean that was to last, in varying degrees, from that day to this.

The nearest sugar island, Cuba, was only a few miles south of Florida, although sugar had been a marginal crop throughout much of Cuba's history. That had changed rapidly in the early nineteenth century with the collapse of sugar in post-revolution Haiti, the emergence of an independent USA freed from restrictive British colonial controls, and the steep decline of British Caribbean sugar in the mid-nineteenth century. In the spirit of free trade, the British had abolished their sugar duties in 1846, and sugar from other regions was now allowed into Britain. British consumers were now able to buy sugar at lower prices, and Cuba, which was already selling half its sugar production to Britain, was ready to meet the increased demand.

Cubans hurried to convert ever more land to sugar cultivation but – once again – the labourers on Cuba's sugar plantations were African slaves. Despite the British and American abolition of the slave trade between 1807 and 1808, Africans continued to be shipped into Cuba and Brazil in unprecedented numbers. In the fifty years following British abolition, more than 500,000 Africans were landed in Cuba.[13] The end result was that consumers in the USA and Britain were still buying slave-grown sugar.

In the last years of the eighteenth century, Cuba had produced a mere 19,000 tons of sugar. Fifty years later, the output reached 446,000 tons from 1,439 plantations, and sugar had become a major driving force behind the transformation of Cuba itself. Its prime market was now the USA and its financial backing was American – but its labour force was African.

Like the sugar planters in Louisiana before the Civil War, Cuba's sugar planters set out to establish a modern, efficient industry, making the most of new machinery, using steam power to run their large, central factories, and operating steam trains to transport the sugar cane and the refined sugar and rum. Mid-century, the island was producing one quarter of the world's sugar output – all made possible by slavery. In the thirty years before the last Africans landed in Cuba in 1867, Cuba received more than 300,000 enslaved Africans. All this was greatly helped by US finance and by US ships – albeit illicitly. Cuba was slowly succumbing to the power of US finance and influence, and US politicians and governments reacted testily to outsiders – notably the abolitionist British – intervening in Cuban affairs. Despite the formal commitment of the USA to abolition, there was money to be made from slave trading to Cuba, and the US took a dim view of British efforts to interdict or board American vessels suspected of slave trading. This ambiguity ended in 1860 with the election of Lincoln, the outbreak of the US Civil War and the ending of slavery in the USA, and of American slave trading to Cuba.[14]

There is, however, a real paradox at the heart of this story. Just when the West – led by Britain and the USA – was formally committed to ending the Atlantic slave trade, and were deploying their navies to stop it, the development of the Cuban sugar economy and the rapid spread of other plantation commodities in Brazil created a new and ravenous appetite for enslaved Africans. Spain, in control of Cuba until 1898, simply turned a blind eye to slave trading, for the simple reason that it seemed good for the Cuban economy. Some years saw upwards of 30,000 Africans land in the island. Over the entire history of the Atlantic slave trade, Cuba received twice as many enslaved Africans as North America.[15] And behind this continuing migration of Africans there lay two interrelated factors – African slave labour and North American consumption. US investment underpinned Cuban sugar plantations and many of the slave ships, and US customers clamoured for ever more slave-grown Cuban sugar.

Cuba's sugar plantations were large and highly mechanised. Cuban tobacco and coffee also thrived, until the take-off of Brazilian slave-grown coffee from the 1840s, and by 1850 Cuba had become one of the world's wealthiest colonies. Yet this rising slave-based wealth was out of kilter with the erosion of slavery across the Americas. Cuban slavery was ultimately undermined by the struggle for independence from Spain. During a destructive decade of warfare between 1868 and 1878, growing numbers of slaves simply voted with their feet and walked away from the sugar plantations. The fighting, especially by ex-slaves, inflicted great damage on the island's sugar regions, and huge numbers of slaves escaped to freedom in the confusion of warfare. It proved ever more difficult to keep slavery in place and Spain reluctantly conceded. Cuban slavery was finally abolished in 1886. Only Brazil clung on longer to slavery, abolishing it in 1888.

Throughout its long history in the Americas, slavery seemed the very foundation of sugar prosperity. Like their predecessors in the French and British islands before them, Cuba's sugar planters found it hard to imagine sugar without slave labour. Now they, too, were forced to look elsewhere for labour. They turned to distant locations. By 1873, more than 150,000 indentured Chinese had been shipped to Cuba; poor Europeans also arrived, alongside the dispossessed poor from other Caribbean islands. But all of them recoiled and fled the harsh conditions of life on the sugar plantations. And although Cuban sugar planters received little comfort or support from their Spanish rulers (officials were corrupt and Spanish politicians unsympathetic), the buoyant market of the USA was close to hand, and was deeply committed to the island via investments and trade. There was even talk, among the wilder fringe of Cuban planters, of joining the US in a more formal political arrangement.

Cuba's sugar trade to Europe had been badly hit by the development of European beet sugar in the late nineteenth century, but the USA came to the rescue, not only buying Cuban sugar but by buying up the island's sugar estates and production via a string of Cuban-American commercial links and family dynasties. Huge American investments and management in large, central sugar factories and new railway systems led to the revival of Cuban sugar by the end of the century. Cuba was now producing a million tons of sugar, most of it destined for the USA, not surprisingly since much of the industry was now in American hands. Even the technical and managerial elite in the industry was American.[16]

By 1896, the US had an estimated $950 million invested in the island, and the US American Sugar Refining Company (1897), which refined most of America's sugar, lobbied to maintain a flow of cheap, raw sugar into their US refineries which dotted the dockside landscape of New York, Philadelphia and Boston.[17]

Cuba was only the most spectacular illustration of what has been called 'America's Sugar Kingdom'. It was a realm forged by US political and corporate power in the Caribbean, and that power was exercised in Cuba, Puerto Rico and, later, in the Dominican Republic. By the late nineteenth century, Cuba had, in effect, become a client state, a fact confirmed by the American invasion of Cuba and the war of 1898. Subsequent legislation secured Cuba as a favoured place for US finance and business. Until the Second World War, the Cuban sugar industry was safely in the pocket of North American financiers, and most of the island's sugar was shipped north to the USA.[18] And herein lies the explanation for the growing American political interest in the Caribbean and in other sugar-production centres. Like Europeans before them, the Americans began to cast a greedy, imperial eye over sugar-producing regions. The Caribbean islands were close, but now the USA also looked across the Pacific.

The Hawaiian Islands had been a base for long-distance whalers, but the decline of that industry, due to the fall in demand for whale oil and whalebone in the 1860s and 1870s, was followed by the emergence of a local sugar industry. An American company had started cane cultivation as early as 1835, helped by missionaries who saw the sugar plantation as a means of converting the local people. After a change in the law in 1850 allowed American investors into the islands, major American companies began to dominate both the land and the sugar industry. The turning point was 1876. On 24 August, the ship *City of San Francisco* docked in Honolulu bringing news of the recently signed Reciprocity Treaty between the USA and Hawaii. It also brought Claus Spreckels, a San Francisco sugar refiner who was keen to take advantage of the new accord. He promptly bought up half of Hawaii's 14,000 tons of sugar for that year, knowing the enormous profits that awaited the islands' sugar in the USA.

After a number of false starts, it was an industry that grew quickly in the 1830s and 40s. In 1836, Hawaii produced four tons of sugar; forty years later, that yield reached 13,000 tons. But the new Reciprocity Treaty, as Spreckels realised, transformed everything. Entry to the voracious US market proved a massive boost to local sugar and, by 1886, Hawaii exported

105,000 tons and Spreckels was lauded (for a while) as the Sugar King in Hawaii. The impact on the islands was immeasurable.[19]

The old Hawaiian kingdom became totally dominated by sugar, which, in turn, was dependent on the American market. Hawaiian plantations and their major American owners and backers wielded enormous economic and political influence in the islands. Like sugar plantations everywhere, the new Hawaiian version needed plenty of labour, but the islands' population was not strong enough to sustain the growing sugar industry. Estimated at between 300,000 and 600,000 at the time of Captain Cook's arrival, it had fallen to less than 60,000 by the 1870s. Many people migrated to the American mainland, especially following the rapid development of California, but sugar badly needed labour. The solution, once again, was to import labour from distant places. Hawaiian sugar planters turned to China and Japan and, by 1890, they had transported 55,000 Chinese and Japanese indentured labourers (charging the Hawaiian monarchy for the system). By the turn of the century, native Hawaiians were a minority in their homeland, and the Japanese were the largest ethnic group in the islands.

What happened to Hawaii's sugar workers was a repeat of an old, familiar tale – harsh working conditions, crude living and social facilities, deep-seated workers' resentment and a persistent, guerrilla-like resistance, marked by regular disputes with sugar planters. Wherever it took root, sugar seemed to breed planters who managed their properties and labour force with a draconian hand; they managed a disaffected and often oppressed labour which bridled against their lot from first to last.

The power of Hawaii's planters also extended far beyond the cane fields. In 1887, a clique of them imposed a new constitution on the Hawaiian monarch, securing planters' control over the government of the islands. A new, gerrymandered political system effectively delivered political power into the hands of Americans, to the exclusion of Hawaiians. Even this was not enough for the sugar men and, following changes in fiscal policy in Washington, and a dynastic change in the Hawaiian Royal Government, the US annexed the islands as part of the 1898 US war against Spain, in the Philippines and Cuba. Hawaii's popular Queen was overthrown by American Marines – acting in the interests of the islands' sugar planters. In 1900, Hawaii became US territory, and the sugar planters seemed to have everything they could wish for. They were now part of the USA and could trade and export freely to and from the mainland. In 1875, Hawaiian sugar accounted for a mere 1 per cent of the US market; by 1900, that had grown to 10 per cent. Hawaii's sugar planters were now securely entrenched in the islands, with a sweet deal with the US mainland, and they were to hold sway until the Second World War.[20]

The planters were alarmed, however, about the rise of the Japanese population – the very population which they had brought into being. They were also unhappy, by becoming American, to lose access to indentured labour, which was now illegal. On the other hand, the Japanese workers, augmented by large numbers of Japanese women, began to organise to defend their own interests. Predictably, the sugar planters responded not merely with their traditional heavy hand, but by the importation of cheap Filipino labourers, who eventually numbered 100,000.

Here, again, was a familiar chapter in the story of sugar, with Hawaiian sugar repeating the experience of sugar in other corners of the world – plantations which drove off indigenous landowners, hard-nosed planters who ran exploitative labour regimes and who came to wield local – and even metropolitan – political power. It was the latest version of international sugar interests whose loyalties lay not at the point of production, but in distant offices and counting houses. Sugar was an industry which seemed to taint all involved, wherever it took root. Worse still, perhaps, the old exploitative relations between master and man on sugar plantations lived on to modern times. The USA was to become home to some of the worst examples.

Hawaii was merely the latest example of sugar's insatiable need for labour on the most exploitative of terms. Slavery had gone, after yielding good returns for the best part of four centuries, and freed slaves simply turned their backs on what they regarded as the house of bondage – the sugar plantation. But what followed was uncomfortably familiar.

In many of the old British colonies, indentured Indian labour took over from the freed slaves. This new diaspora of indentured labour was to fill the gaps left by the departing freed slaves, or to work on new lands and in new colonies such as Guyana and Trinidad. It was, again, a massive movement of humanity which continued into the early twentieth century.

Moreover, this Indian diaspora to European colonies was not restricted to the old slave colonies of the Caribbean. In the Indian Ocean, Mauritius absorbed almost half a million Indians; Reunion, 87,000; Natal in South Africa, 152,000; and Malaya, 250,000.[21] Sugar was, yet again, the main driving force behind these migrations, much as it had been with slavery. It was, once more, responsible for transforming the demography of parts of the globe.

The use of indentured labour in sugar was successful and offered a blueprint for other crops, and which led to further major migrations – indentured Chinese and Japanese labour were moved to the Caribbean, South America, Hawaii and California. Here was a supply of cheap alien labour that could be tied down to a particular location or crop – it might be sugar or pineapples, tea, palm oil or, later, rubber. Labourers were tied to an employer for a specific number of years in return for certain terms of employment. It certainly was not the same as African slavery. Nonetheless, indentured labourers were less than free. This shift from slavery to indentured servitude by the colonial powers looked, to many critics, a classic example of imperial humbug. Countries proclaimed their virtue in ending slavery – none louder than the British – but continued to consume huge volumes of sugar produced by people who had been shipped vast distances to work in oppressive conditions. Long after slavery had ended, sugar continued to provide sweet pleasure for millions at the cost of exploitative conditions for its labour forces.

In the last twenty years of the nineteenth century, Cuba became the world's largest sugar producer. When slavery there was abolished in 1886, the island was producing 750,000 tons of sugar, representing around 40 per cent of the world's exports. And it did so with finance from Spain, the USA and Britain, and using modern equipment shipped across the Atlantic from British heavy-engineering companies. But behind this story of Cuban sugar lay the rising power of the USA. It had sustained Cuban slavery by shipping in Africans on American ships, it invested heavily in Cuban plantations, and the US bought much of the island's sugar exports. Now, after 1886, Cuban sugar planters faced the problem that had confronted earlier planters elsewhere in the Caribbean – how to run sugar plantations without slaves?

In the years after the abolition of slavery, Cuban sugar came to dominate Cuban agriculture. It greatly surpassed the output of local coffee, with Brazilian coffee now dominating world markets, i.e. mainly the USA. By the turn of the century, Cuba was producing more than 1 million tons of sugar, most of which went to the USA – not surprisingly, since most of Cuba's sugar industry was owned, controlled, directed or managed by US and joint US–Cuban interests. (In 1885, for instance, some 250 skilled men from Boston alone worked on Cuban estates.) The link was simple, as the US Consul in Havana admitted: 'De facto, Cuba is already inside the commercial union of the United States.' The American Sugar Refining Company (the Sugar Trust), which refined 70–80 per cent of America's sugar, owned nineteen refineries in Cuba. It wielded enormous economic and political power in Havana and Washington, and played a critical role before and after the 1898 Spanish–American War, ensuring that Cuba remained securely within the gravitational pull of Uncle Sam.[22]

Cuba was Uncle Sam's tropical backyard. It helped to feed the US demand for sugar, and forged that special relationship so defended by the US, yet so disliked by Cubans, and which became central to the spiky and often belligerent relations between the two countries from that day to this. At the heart of this persistent rumbling discord lay the matter of sugar.

The US interest in Caribbean sugar extended far beyond Cuba. The three Spanish islands of Puerto Rico, the Dominican Republic and especially Cuba provided rich pickings for US interests. Via an astonishing and unprecedented drive to convert ever more land and labour to sugar cultivation, they were producing 1.3 million tons by 1902, before doubling it again by 1920. Three years later, their output had risen to an astonishing 6.75 million tons.[23]

Worldwide sugar prices rose steeply in the early years of the twentieth century and, as a result, the old sugar regions expanded, and new regions rushed into sugar production: Mauritania, Brazil, Argentina and India, for example. The American appetite for sugar was now sated by imports from Hawaii, the Philippines, Puerto Rico and, above all, Cuba. All this in addition to the home-made sugar from Louisiana and from the sugar-beet industry in the US Midwest, and from Germany. But a reliance on imports remained vulnerable to disruption. Just how vulnerable was starkly illustrated by the outbreak of the First World War.

America's ability to produce enormous volumes of food-stuffs proved vital in the First World War, but wartime losses of shipping threatened food supplies on both sides of the Atlantic. America's imports of beet sugar from Germany simply dried up and it was clear that the market needed more sugar from the Caribbean. But it soon became apparent that the USA also needed to ration key food items and to persuade Americans to change their eating habits. Sugar was an obvious target and the US Government set out to persuade people to consume less sugar, especially when the USA entered the war in 1917. Sugar was now required to supply the US military and allies in Europe.[24]

The Food Administration, given the task of handling the problem, focused on wheat, meat, fats – and sugar. There was a pressing need to reduce America's per capita consumption of sugar (currently running at 90lb), and hotels, clubs, restaurants and stores were all instructed to reduce the volumes of sugar in their fare. Manufacturers of candies, soft drinks, chewing gum and ice cream – all were expected to reduce their sugar contents.[25]

One outcome of wartime restrictions was an agreement between the US Government and US sugar refiners to create a monopoly in sugar buying, with a joint US–European committee deciding how best to divide up sugar between the US and her European allies. To many Americans, the resulting sugar rationing, and the constant prompting to be self-denying and parsimonious in sugar consumption, smacked too much of government interference in people's private lives. The altruism of such pleas (for example, that the volumes of sugar used to manufacture US candies was enough to cover the entire sugar consumption of Britain and France) did not please or satisfy everyone. Major sugar refiners were equally quick to bridle at government controls. The result was a widespread unpopularity of 'big business' (which had its roots in the anti-trust mood of a generation earlier). The Sugar Trust was especially disliked; it seemed more concerned for its profits than about national interest, or low prices for the American consumer.

America's wartime sugar policy focused on luxury items – candies, chocolates, ice creams – although exceptions were made to encourage home preservation of foods and fruits, by allowing extra sugar.[26] By the summer of 1918, however, the war had created severe sugar shortages. Millions of pounds of sugar had been sunk by German submarines, and President Wilson authorised the establishment of a sugar corporation, empowered to import sugar and control the profits on the handling and sale of sugar. Restrictions were placed on the use of sugar in public and even in the home. Consumers were now limited to one ounce a day, with certificates introduced to control and monitor the whole process. Rationing was tightened and newspapers were recruited to keep up the pressure with exhortations to reduce consumption: 'SEVEN WAYS FOR

DAILY SUGAR SAVING'. Open sugar bowls were banned in restaurants. Despite the difficulties and criticisms, the system worked. Americans had been asked to save 600,000 tons of sugar. In the event, they saved 775,000 tons.[27]

America's problem with sugar in wartime was only the latest example of a story that had been familiar for the past three centuries. Sugar, produced primarily in distant tropical places, had become an everyday necessity in the lives of millions. But it was vulnerable to threat of disruption in wartime. To keep supplies moving, and to keep prices acceptable to consumers, government was forced to intervene and, in 1917, as in the eighteenth century, sugar was a hotly contested issue. It had so intruded into the diet of millions of people that life was unimaginable without it. Sugar was essential – and it was everywhere. If it was scarce or costly, politicians and governments were called to account.

The high sugar prices in wartime were bound to fade in peacetime, and although they held up at first, by the mid-1920s they began to fall steeply. After the Wall Street Crash of 1929, economic and social discord abounded. In Cuba, the impact of the crash on the dominant sugar industry was catastrophic, with unrest, revolution and the physical takeover of American-owned sugar plantations and mills. US sugar growers – powerful in Washington – demanded, and got, tariffs on cheap, imported sugar to protect them from Cuban sugar. FDR, President from 1933, worrying about the very real danger of upheaval in Cuba, produced a compromise – an equal quota in the US market for both US sugar and for imported foreign sugar, with Cuba granted 64 per cent of the foreign quota. The Sugar Act of 1937, which remained in place until 1974, 'made sugar the most regimented of all American crops'.[28]

On the eve of the Second World War, sugar had become a remarkably regulated and controlled commodity in the USA. Within a generation, sugar had seduced the US Government to involve itself in the most intimate and detailed matters of the sugar industry. In living memory, sugar had been at the heart of US foreign (and military) politics in the Caribbean and Pacific, and had been critical in pushing the US Government into detailed intervention in the economy. What underpinned this entire story was the American consumers' attachment to sweet food and drink. Americans like to use the phrase 'as American as apple pie'. That pie needed plenty of sugar to make it tasty.

No one doubted that sugar was 'a necessity of life . . .' in the USA.[29] Moreover, much of it came from Cuba and, after 1945, that island enjoyed a lavish quota in the US market. All this was thrown into confusion by the Cuban revolution of 1958. Fidel Castro set about tackling the US-dominated sugar industry on the island. He confiscated sugar estates of more than 1,000 acres and banned foreigners from owning Cuban land. The US Congress promptly slashed Cuba's sugar quota, and Cuba turned to new trading partners in Russia and Eastern Europe. When President Kennedy banned all trade with Cuba, the USA immediately lost half of its sugar supplies. Faced by this enormous shortage, Congress promptly granted Cuba's sugar quota to American producers, lifting the 26-year-old restrictions on US sugar-cane production. Quite simply, American growers were given 'the green light to plant as much sugar as they could grow'.[30]

In one state in particular, the prospects were too tempting to resist. Major agribusinesses swooped on the most obvious location – Florida – for an expansion of sugar cultivation. Florida already had a fledgling sugar industry. It had produced a mere 29,000 tons in 1934, which increased to 175,000 by the 1950s. Compare that, however, to Cuba's mighty output which was in the millions of tons. The centre of Florida's sugar industry, the small town of Clewiston – 'America's Sweetest Town' – now became the focus for a massive expansion into surrounding sugar fields. New sugar lands, however, required heavy investment both in land and in machinery. Factories which cost millions of dollars to build also required a steady flow of sugar cane to function smoothly. Many Florida agriculturists specialising in other types of produce – fruit, vegetables, cattle – transformed themselves into sugar planters, and new co-operatives of sugar growers sprang up, producing more and more sugar for the American market.

The most important newcomers to the post-1958 Florida sugar boom were exiled sugar planters from Cuba, some with vast, lucrative sugar businesses and family possessions in Cuba seized by Castro. In Florida, they picked up where they had left off in Cuba, buying land, machinery and investing heavily in establishing what was, effectively, a new sugar industry. By the 1990s, they had more than 190,000 acres in sugar cane. Cubans built eight major factories and employed legions of émigré Cuban workers. Americans with experience of Caribbean sugar – and with access to US finance and a desire to oust Castro – also joined in this Florida 'sugar rush'. By the end of the twentieth century, the US Sugar Corporation in Clewiston was producing 700,000 tons of sugar annually. This rush into sugar, however, had serious and complex consequences for the Florida Everglades which bordered their sugar lands.[31]

In the space of thirty years, yet another new breed of sugar baron had emerged. They, too, were wealthy beyond imagination, controlled huge landholdings in Florida and the

Caribbean, and they owned highly modernised factories disgorging large volumes of sugar for the US market – all at a price guaranteed by the US Government. Florida's sugar men were also very well placed, with friends in high places in Washington on whom they lavished costly hospitality and financial help in elections.

The consequences for the state of Florida were enormous. In 1955, the state had 36,000 acres turned over to sugar cane; by 1973, that had risen to 276,000 acres. Florida's sugar output of 173,000 tons in 1953 had grown to almost 1 million tons by the mid-1960s. And the inevitable dark side to this story? It was yet another dismal chapter in the history of sugar, with the Florida sugar industry needing plenty of field labour. And that meant temporary migrant labour, especially from Mexico and the Caribbean. To many, it looked uncomfortably like a new slave trade.[32]

Migrant labour became a global phenomenon in the twentieth century, and US agriculture became particularly reliant on it. Its immediate origins lay in the Second World War and, in a string of agreements between the US and Mexican and British governments to provide labour for the USA, Mexicans arrived by train and by bus, Bahamians and Jamaicans were shipped, and later were flown in, and the patterns continued in peacetime. By the mid-1960s, half a million Mexicans a year were travelling to the USA to work in agriculture. In the same period, upwards of 20,000 guest workers – mainly West Indians – arrived to cut Florida's sugar cane for 5–6 months each year.[33] On the sugar estates, their living and working conditions were dire – bunks in barracks, unventilated buildings, poor sanitation and almost non-existent healthcare. Their employers spun a fine web of propaganda about the facilities, promoting a benign image of life and work in the Florida sugar industry. One film intoned: 'To watch a West Indian wield a cane knife is to see a centuries old art.'[34] This risible remark overlooked the obvious fact that it was an art fashioned by slavery and indentured labour.

The sugar estates tried to shield their employees from outside scrutiny, but investigative journalists were quick to expose the realities, both of the workers' plight and the ecological degradation wrought by the expansion of the sugar land on to the Everglades. Sugar's exhaustive use of scarce water supplies, and the poisonous run-offs of waste and chemicals into the natural habitat, created serious damage.[35]

The need to preserve the Everglades had been obvious in the early years of the twentieth century, although the area was not designated a National Park until 1947. Urban growth and industrial development had profound effects by mid-century, although most damaging of all was the expansion of sugar cultivation. Once again, sugar seemed to be the catalyst for the degradation of a precious national habitat, and prompted an environmentalist (eventually a political) movement to return the sugar lands back to their natural state. At the time of writing, negotiations for the state to buy back the sugar lands for $1.7 billion have stalled.

Within a generation, prompted by a revolution in Cuba and by the survival of a protected US sugar market (put in place in wartime), Florida's sugar industry had emerged, alongside tourism, as one of the state's major industries. The consequences for labour, and for the delicate ecosystem of the Florida Everglades, were enormous. Sugar was, yet again, living up to its reputation as a threat to its natural habitat, an employer on inhuman terms and – as we shall see – a product with catastrophic consequences for the health of its consumers.

The recent story of sugar in Florida was, in many respects, an updated version of the history of European colonial sugar two centuries earlier. Sugar cultivation had transformed the ecologies of the Caribbean in the seventeenth century, and threatened to do the same in the Florida Everglades in the late twentieth century. Although clearly not on a comparative level with slavery, Florida's sugar industry also generated a wretched story of labour exploitation – it was brutal, callous and degrading. The conditions of transient workers in Florida's cane fields prompted a string of legal cases. Importing migrant labour was a good way of circumventing local labour laws; migrants could be moved on at the first sign of discontent or complaint. In the worst cases, in the 1980s, police were used to break up – and beat up – striking workers, before dispatching them back to their homelands.

Yet at the same time, astonishing wealth was generated by the system. The twentieth-century Florida sugar planters lived in a style, and boasted wealth and assets, that would have shamed the sugar barons of the eighteenth century. Equally, the wealth of Florida's major sugar barons enabled them to court, and be courted by, politicians. Two hundred years earlier, planters, like Indian nabobs, had exercised political influence in their homelands in defence of their sugar interests. So, too, the barons of Florida. They became important players in American politics, both at regional and state level. Not surprisingly, the ecologists' efforts to curb the dangers posed to the Everglades faced enormous political obstacles. They still do. Although the rise of the modern environmentalist movement has opened the public's eyes to the damage caused by the intrusion of sugar – as well as urban development – into the Everglades, their efforts have only partially succeeded. As with the wider story of sugar itself, those who have the most to profit from sugar – the people who cultivate, process and market it – show few signs of easing up in their drive to ever-greater production.

Throughout the twentieth century, American governments of all persuasions had taken a keen interest in sugar. But why was this so? What was so special about sugar that prompted governmental interest and intervention? To an outsider, it seems odd that this one tropical commodity should become a matter of sensitive political and even strategic concern. Yet this had been the case with Europe's sugar economy between the seventeenth and nineteenth centuries. And it was true of the USA in the twentieth century.

From the early days of the Republic, the US Government had played a critical role in the story of sugar in the USA. In the century between 1789 and 1891, the USA levied a tariff on imported sugar to raise revenue for the Treasury. Import duties provided two thirds of US Government income in the nineteenth century, and sugar duties made up a hefty 20 per cent of those duties. Higher duties on imported refined sugar protected US sugar refiners, while tariffs on raw sugar were designed to protect US cane growers. After 1890, new tariffs had the desired effect of stimulating the expansion of the US domestic sugar industry. Thereafter, Cuban sugar poured into the refineries on the east coast of the USA and sugar (and its tariffs) entered the bloodstream of US politics, law, and even of global politics. America's sugar politics had major consequences on the economic and political stability of Hawaii and Cuba, in both places bringing about conflict and the eventual imposition of direct American control. Clearly, the diplomatic and economic turbulence which led to war in

1898 was a complex brew. But sugar was a decisive element in that mix.

After the 1898 US acquisition (although sometimes only temporary) of new overseas territories such as the Philippines, Hawaii, Cuba and Puerto Rico, the USA had come of age as a major international and imperial power. Curiously, sugar had been a major consideration in the way the USA tried to balance its domestic and global interests and obligations. In 1900, no other foodstuff had become so important in the USA and its international dealings.

By the early twentieth century, sugar had become vital to the US economy as a whole, not merely to the vested interests of refiners and manufacturers. It was an integral feature of economic and social life, and any threat to supplies – or to the price – of sugar could cause serious disruption to American life – with untold political consequences. This was confirmed in spectacular fashion, both in Europe and America, during the course of the First World War.

Today, it seems odd – far-fetched even – to suggest that sugar could have proved so instrumental an issue in international matters. Yet contemporaries were in no doubt. Arguments about sugar were not reserved for specialist journals or cabals of vested interests. The 'sugar question' was a deeply contentious issue with far-reaching international consequences. Sugar was not simply a domestic matter – it was an explicitly international problem concerning the rights of new territories and the responsibility of colonial power. In 1897, for example, the *American Monthly Review of Reviews* published an article entitled 'Sugar – The American Question of the Day'.[36]

In 1900, sugar was widely discussed in newspapers and magazines. It was a source of political controversy and of economic theorising; it was a matter of political skullduggery and high diplomatic principle. Much the same was true a century later. Cuba and the USA entered the twenty-first century overburdened by a mutual historical legacy that had been conceived and nurtured in the story of sugar.

12

A Sweeter War and Peace

FOR CENTURIES, EUROPE'S slave empires had ensured that their metropolitan hinterlands would be avid consumers of sugar. When slavery vanished (in the British case, in a puff of self-righteous celebration), it left behind an unquenchable appetite for sweetness in all things. That taste continued to be sated by a changing mix of cane sugar from former slave colonies, from new tropical settlements, and from beet sugar yielded by Europe and America's expanding beet industries. The British imported large volumes of German beet sugar, but that came to an immediate end with the outbreak of war in 1914.[1]

Sugar was big business in Britain itself, its importance visible enough in the form of refineries in many British port cities. London had been the home to Britain's first sugar refineries, but the rapid expansion of the industry saw the proliferation of refineries at many of the country's major ports – Liverpool, Bristol and Greenock – though the majority were in London. In 1851, for example, there were an estimated 1,200 'sugar refiners' working in the city.[2] But it was an industry which, like most others, changed in the nineteenth century with the rise of modern technology, which speeded up the process of refining cheap granulated sugar. Where it had once taken weeks to produce sugar loaves, now tons of granulated sugar spilled from London's modern refineries in a matter of days.[3] New machines also made possible the mass manufacture of a host of new sweets and chocolates. What had once been slow manual labour was now speedily dispatched by machines, as described by Henry Weatherley, a machine manufacturer, in 1865:

The large increase in the consumption of sweets, made from boiled sugars, in the UK during the last quarter of a century, has arisen principally from the cheapness and facility of manufacture, derived from the introduction of machinery . . .

Where it had once taken half an hour to make boiled sweets by hand, a machine now took a mere five minutes.[4]

Along with Hamburg, London had become a major trading centre for global sugar. The East End of London had been dotted with sugar refineries, mainly strung out along the river in Whitechapel and along the Commercial Road,[5] but their numbers declined with the advance of sugar technology. In 1864, seventy-two British refineries processed 500,000 tons of sugar; by 1913 this had been reduced to thirteen refineries, but they were producing more than 1 million tons of sugar.[6] The process of technological and scientific change transformed the food habits of Europe and North America after *c.*1850 and, as a result, millions of people turned to a new urban diet, all of which was discharged by modern machinery and delivered swiftly by new transport systems. Moreover, wherever we look, cheap granulated sugar lay at the heart of those industrial foodstuffs.[7]

Cheaply produced sugar and, from the mid-century, an era of free trade enabled sugar to consolidate its position as a central ingredient in the British diet. In 1810, the annual per capita consumption of sugar in the UK was 18lb. That had doubled by 1850 to about 30lb, rising ever further as the century advanced: 68lb in the 1880s; 85lb between 1900 and 1909; and a truly astonishing 91lb on the eve of the First World War – almost half as much again as the German average.[8] By the mid-twentieth century, that had increased even further to 110lb. By this time, despite the staggering volumes of sugar being consumed, plant modernisation and corporate amalgamations actually saw the number of British sugar refineries decline, from thirteen in 1900 to seven in the 1970s.

Such enormous levels of sugar consumption, and the related levels of imports of cane and beet sugars, inevitably caught the eye of governments. Arguments about duties, tariffs, about protection for colonial sugars, and/or free trade, all ensured that sugar was as important a political issue in the twentieth century as it had been in the eighteenth. In both Britain and in the USA, sugar was basic to the nation's diet and drink, guaranteeing that it remained at the heart of heated political debate in both Parliament and Congress.[9]

The British were in good company in their attachment to sugar. The Western world at large was thoroughly dependent on sugar by 1900. It was consumed in great volumes by all levels of society, in all corners of the world – clean across North America, in the expanding communities of Australia, and in every major country of Western Europe.

Australians led the way, devouring 107lb each in 1900, with the British close behind. This remarkable story, rooted in a colonial past, was transformed by the emergence of beet sugar which, by the mid-nineteenth century, had replaced cane sugar as the principal source of sweetener in Europe. Even in Britain at the beginning of the twentieth century, 80 per cent of Britain's sugar came from European beet – and most of that came from Germany and Austria, whose imports had been greatly encouraged by the British abolition of duties on sugar imports. The result was that sugar was much cheaper in Britain than in Europe.[10]

The increasingly urban and industrial world of the mid-nineteenth century enhanced the position of sugar. In fact, modern industrial societies devoured sugar as never before. Looking back from 1936, an agricultural scholar claimed that the rise of sugar consumption was 'the most significant change in the nation's diet during the last hundred years'.[11] This had come about partly because of the fall in sugar prices caused largely by the removal of duties, by the expansion of sugar production, and by the massive increased use of sugar in the industrial production of basic foodstuffs. Sugar became a major additive to other foods and drinks as it had been with tea and coffee for centuries. People did not consume sugar on its own – they added it to other foods and drinks. Whatever the British bought from their shops and corner stores – tea, bread, flour, bacon, jam – they also bought sugar, which, in the words of Peter Mathias, was 'the greatest complementary commodity . . .'[12] Now, however, manufacturers were also adding sugar to food *before* it reached the customer. Sugar was in the flour that made the nation's bread, it was in the beer they quaffed by the barrel load, and it was added, in industrial volumes, to cakes, biscuits, chocolates and sweets. Above all, perhaps, sugar was poured into the astonishing range of jams which became a central feature of the diet of British people – especially poorer people. Jam factories sprouted throughout urban Britain, normally employing cheap female labour. Jam companies evolved from fruit farms, which made the most of new canning and bottling facilities, and branched into jam manufacture in the late 1800s. Chivers and Sons, for example, grew out of an East Anglican fruit farm to become one of the country's major jam producers. Marmalade – most famously perhaps Keiller's of Dundee – grew almost by accident from a decision to convert unsellable Seville oranges into marmalade in 1864. They, too, used the new canning and bottling systems.[13] Lipton's opened their first shop in Glasgow in 1871; by 1914, they had 500 across Britain, all aiming at the working-class customer. One of their best-selling products was jam, which was made in a huge range of flavours in their own factories with the fruit grown on their own farms. But it was always manufactured with the copious addition of sugar.

Lipton's jam factory in Bermondsey in 1892 churned out jams in enormous volumes, in a range of flavours and weights, and all promoted via aggressive, eye-catching advertising and promotion.[14] The end result was that working-class diets, by the late nineteenth century, particularly for women and children, seemed to rely on sugar – highly sweetened tea, jam and bread. Social investigations of the urban poor from the East End of London to the slums of York confirmed, time and again, the importance of sugar and jam among the very poorest of British people. Millions of people found their basic sustenance in sweet jams spread on bread made from flour with added sugar, and all washed down with sweet tea. Not surprisingly perhaps, the emergent dental profession found abundant evidence of dental decay among working-class children in the late nineteenth century. For all that, doctors accepted that sugar was an important provider of energy for labouring people: 'Its real value as a muscle-force-producing substance is not fully recognized.'[15]

For people with a little extra cash to spend, there were plenty of new foodstuffs to enjoy – although they were always accompanied by vast amounts of sugar in new industrial foods. Cakes and biscuits, sweets, chocolates, jams and syrups – all and more flew from the new industrial plants of the food industry, tempting the British consumer from the shelves of new retail outlets. Breakfast cereals (invented by Dr Kellogg in the USA in 1899 as a 'health food') quickly established themselves as an essential feature of the breakfast table. And once there, they were lavishly drenched with sugar. By 1912, the British could choose from sixty brands of breakfast cereal.[16] But of all the new foods to establish themselves before 1914, few could match the impact and scale of confectionary products.

These years were most notable perhaps for the remarkable success of the major British chocolate companies – Fry, Cadbury and Rowntree. They blossomed via their modern industrial and technical production lines and distribution systems – Rowntree built their own railway line to link the factory to the mainline – and all operated in large factories. They also housed workers in neighbouring 'model villages'. Almost everything they produced – chocolates and sweets – contained huge quantities of sugar. Around 1914, all three companies had a turnover in excess of £1 million, and images of their best-selling sweets and chocolates were festooned on billboards, in newspapers, on the sides of buildings and on buses up and down the land.[17] Throughout the war that followed, just as in the Boer War, the chocolate magnates ensured that troops in the trenches received regular supplies of their favourite chocolates and sweets.

Alongside the chocolate manufacturers, the years before the First World War also saw the rise of major British manufacturers of cakes and biscuits. Many are familiar because their names survive to this day, although their corporate identity has often been swallowed by global conglomerates. These were the years when the British became a nation of biscuit lovers. In 1900, Huntley and Palmer manufactured 400 varieties of biscuits; Peek Freans had 200 varieties. They were stocked on grocers' shelves alongside an array of jams, Lyle's Golden Syrup, tins of highly sweetened condensed milk, and chocolates and sweets – all of which had been manufactured in large, modern factories. All these products, and many more, were cheap – and were filled with sugar.

They also accounted for a massive share of the nation's sugar consumption – an estimated 11oz of sugar were consumed by the average Briton each week.[18] By 1938, for example, British homes consumed 1.1 million tons of sugar directly, but another 300,000 tons went into the manufacture of confectionary products.[19] In the years between 1880 and 1914, when the expansive confectionary industry disgorged their biscuits, cakes, sweets and chocolates by the ton, the British per capita consumption of sugar increased by more than one third from 68lb to 90lb.[20] The British were consuming vast amounts of sugar in their sweets, chocolates, cakes and biscuits.

The outbreak of war in 1914 transformed everything. So vital was sugar that, within days of war being declared, the British Government established a Royal Commission for Sugar

Supplies. It was to mark a quite remarkable phenomenon, and one that was to last throughout much of the twentieth century – state involvement in the procurement, distribution and pricing of sugar. Here, at a stroke, was the most telling indication of the commanding position of sugar in the diet, politics and economy of the British people.[21] Of course, this was only one aspect of the much broader story of wartime state intervention into most corners of the social and economic life of the nation. To fight the war that was declared in 1914 required the state's involvement in, and control of, all sorts of commercial activity which would have been unthinkable in peacetime. Looking back, it is now clear that such state intervention – which was even more dominant and pervasive after 1939 – would not be easily removed. The state was here to stay and, among many other issues, the state thought it vital to manage the nation's sugar.

The British Government responded to the loss of European beet sugar supplies by scouring the world for other sources – from Java, Mauritius, Cuba and other Caribbean islands. Sugar rationing was inevitable, and remained a working-class complaint throughout the war.[22] And while the restrictions of wartime proved upsetting for consumers, government assistance and subsidy proved a boon to the domestic beet growers and refiners. For the next sixty years, the British sugar industry was to thrive on the back of state support.[23]

The enormous British military machine of 1914–1918 swallowed up huge quantities of sugar to feed its soldiers and sailors. Major confectioners – of jam and biscuits, of chocolates and sweets – were kept busy despite sugar rationing, providing the millions of men in uniform with the sweet essentials they had grown accustomed to in peacetime. Jam, rum and sugar figured prominently in the military's basic foodstuffs. When Robert Graves left his public school to join the Army in 1914, his first meal at the front consisted of 'bread, bacon, rum and bitter-stewed tea sickly with sugar'.[24] There was sugar in the tea, sugar in the bread and rum extracted from sugar cane; and rum was vital on the eve of combat, that gulp of alcohol giving a much needed boost to men about to 'go over the top', lending an iota of steely resolve in the face of the horrors to come.

Wartime sugar shortages taught the British a lesson. The immediate post-war years proved a boom time for sugar refiners, as it did for their US counterparts. Anxious to learn the lessons of wartime, the British governments from 1921 began a vigorous promotion of the British beet industry to avert the problems caused by the earlier reliance on European beet. At the same time, they increased imports of cane sugar from imperial sources. Sugar began to arrive from new sources – Australia, South Africa and Fiji – and although the British beet industry went through hard times after the war, it survived in large part because of government help. The end result was a massive over-production of sugar, and a consequent political and economic balancing act between producers, refiners and government officials. In effect, the British sugar industry was subsidised by the British Government; taxpayers' money went into the pockets of sugar-beet farmers. Unwittingly and unpredictably, sugar beet had become essential both to British agriculture and to the British consumer. Thanks to the taxpayer, beet farmers and refiners now showered the British people with unprecedented volumes of sugar. In the process, the major sugar refiners emerged as a real power in the land, acquiring substantial wealth and influence, and insinuating themselves into a range of commercial outlets.

Until rationing intervened in the Second World War, British sugar consumption simply grew and grew – by 50 per cent between 1880 and 1939. In the century to 1936, there had been a fivefold increase in British sugar consumption. These were the years of falling sugar prices, and the years when sugar established itself – along with wheat and potatoes – as a key source of carbohydrate in the British diet.

Although working people relied on sugar for the energy needed in their daily labours, by the late 1930s British sugar consumption was fairly evenly distributed among all social classes. It also remained an important ingredient throughout Britain's food manufacturing industry. Indeed, by then about 40 per cent of the sugar consumed by the British was in industrially processed foods. The list of those foods containing sugar is so familiar because it represents a modern-day problem, from chocolates to breakfast cereals.[25]

What added a new dimension to the entire question of diet was the development, in the early twentieth century, of the modern science of nutrition – the discovery of vitamins, amino acids and mineral elements. These provided scientists and medical officers with the means of studying the nature and quality of the nation's nutrition. At the same time, there was a continuing social scrutiny and analysis of the extent and levels of poverty, and of physical well-being in the British population. From Charles Booth in London, to Seebohm Rowntree in York, through to the initial steps towards a welfare state guided by Lloyd George and Churchill before 1914, a vast body of literature had emerged about the nature and causes of British poverty. It was now widely accepted that between a quarter and one third of the urban population was poor, and existed on an inadequate diet. Bit by bit, the new nutritional sciences confirmed that such problems could not be solved merely by giving people *more* to eat; they needed different and better food.

The entire problem came into sharper focus via compulsory schooling which brought every child under the scrutiny of medical and dental experts. What they discovered was alarming. Beginning in the 1880s, the cumulative findings of social and medical investigators made it clear that the British poor (which included millions of unemployed between the wars) needed better food – they required 'milk, fresh vegetables, meat, fish and fruit'. Such arguments generally fell on deaf political ears at a time when government finances were stretched and, more important perhaps, when the political will was lacking. It was to take the Second World War, with its far-reaching state intervention into citizens' lives, followed by the creation of the modern welfare state in the late 1940s, before the lessons of nutritional well-being really began to make an impact among the British people.

The development of scientific analyses of food also raised troubling questions both about sugar in the British diet and, more disconcertingly, about the very nature of sugar itself. A number of studies revealed sugar to be 'wholly devoid of minerals and vitamins'. It had also become clear that large-scale consumption of sugar was having a damaging effect on children's teeth. In the years when industrialised, sugary foods and drinks came to dominate the national diet, it was evident that the nation's teeth were being corroded by sugar.[26]

* * *

The return of warfare in 1939, and the draconian though vital food rationing throughout the conflict, once again reinforced the importance of sugar in the British diet. Along with a long list of other imported foods, the main problem was supply. European sugar beet was again under German control, and imported cane sugar from tropical sources faced the enormous dangers of submarine attacks on allied convoys. Sugar was immediately rationed to 12oz per person per week (today, that seems an abundance), and food manufacturers were limited to an allocated percentage of their pre-war consumption. The British sugar-beet industry was encouraged to expand, and intricate financial arrangements developed between farmers, refiners and government. What evolved was a national sugar industry controlled by a complex political and financial arrangement to regulate the production and sale of sugar. It was to last long after the war was over.[27]

Sugar rationing began in January 1940 and lasted, with one brief interlude, until 1953. An Act of 1942 put the nationwide system in the hands of a subdivision of the Ministry of Food and, as wartime hardship began to bite, when German submarines seemed to be winning the 'Battle of the Atlantic', rations were cut further. Private diaries captured the mood. Kathleen Hey, a shop assistant in Dewsbury, noted in her wartime diary:

Sounds of prolonged groans because the rations are cut, particularly sugar, which seems even more than tea to be the thing people would like more of.

It was widely agreed that men took the lion's share of sugar, having on average two to three spoonfuls for each cup of tea. Men, she thought, 'are the sugar devourers'.[28]

Government control of sugar – working hand in hand with Tate and Lyle, by then the dominant force in the British sugar industry – was part of a remarkably intrusive state system which was to prove very difficult to dismantle when peace returned in 1945. The British taste for chocolates and sweets, for example, was so voracious and entrenched that when rationing of those items was removed in 1949, demand was so immediate and widespread that rationing had to be re-introduced for another four years. The irony was that the British emerged from the war even *more* wedded to sugary foods than before 1939. Sugar consumption continued to rise, peaking at 115lb (52.4kg) per capita in 1958. Although the figures declined thereafter, they remained at a remarkable 40kg and more a head as late as 1990.[29] The British entered the post-war era of rising material prosperity deeply attached to sweet food and drink.

The British sugar industry was now dominated by Tate and Lyle, which had been formed in 1921 from two earlier business competitors. The sugar industry was so central a feature of British life that it was placed on the list of essential industries designated for nationalisation by the post-war Labour Government. The fact that Labour should even *consider* nationalising sugar, alongside railways, coal, steel and health, is the clearest possible indicator of the importance of sugar in British society at large. Here was an industry that had insinuated itself into the very core of the British way of life – the British people seemed not to be able to function without their regular intake of sugar. That intake had increasingly come from a huge food and drink industry, which had developed its own dependence on sugar. Attlee's Labour Government thought it appropriate to take sugar into national ownership, but Tate and Lyle and its shareholders naturally resisted, launching a powerful campaign of publicity and propaganda – 'Tate Not State' – to fend off state control. The company was radically restructured to shield the shareholders' investment from the Government's grasp. 'Mr Cube', the clever icon of Tate and Lyle's campaign, remained a potent commercial image long after the Government's attempts at nationalisation had been rebuffed.[30]

There is, however, an abiding irony to this story. The British sugar industry had been shored up and defended by state protection and subsidy; indeed, it had been utterly reliant on it. Throughout the twentieth century, sugar in Britain had been an industry which thrived, in peace and throughout two world wars, on the paternal involvement of the British state. By securing supplies of sugar from overseas, by providing financial encouragement for domestic beet growers, and creating Treasury-backed deals to safeguard the sugar-refining industry, the state had been the careful guardian and protector of the people's appetite for sugar. Yet Mr Cube was now deployed – his face festooned the sides of buses, placards, advertisements in newspapers and bags of sugar – to resist the very state that had been its saviour. Without the attentive interest and financial support of the British state, sugar could never have maintained its role as a vital industry at the heart of British eating and drinking habits.[31]

Wartime support for sugar continued in peacetime using wartime experience as a guideline. Agreements were struck with Commonwealth sugar producers, refiners were allocated guaranteed volumes of cane and beet sugar, and British farmers producing beet were given guaranteed quotas and prices. A Government 'Sugar Board' managed the entire system, with a minimum of staff and fuss. But all this began to change from the 1970s, with beet replacing sugar cane as the main supplier of sugar. Even greater change was brought about by the profound impact of British entry to the European Economic Community, later to become the EU.

At the end of the twentieth century, the British sweet tooth was less pronounced, but still astonishingly insistent. And sugar was an inescapable feature of British social life. But the political arrangements – the global deals – between companies and governments were about to change by Britain's belated entry to the European Common Market. The industry had entered into a process of complex international deals and negotiations that were so Byzantine, so complex and confusing, that they defied simple explanation. And all this for a product that was to become a cause of great medical concern by the end of the century.

13

Obesity Matters

Over the past generation, there has been rising concern about the level and the apparently relentless increase of obesity. The term is often used loosely to describe those who are significantly overweight with a high proportion of body fat, but the commonly accepted medical definition refers to anyone with a body mass index of 30 or more. In 2015, the World Health Organization (WHO) estimated that around 2 billion of the world's population – nearly a third – were classified as overweight, of whom around 600 million would be clinically recognized as obese. And, even more worryingly, that is a figure that has doubled since 1980.

Obesity is not restricted to the West; there are millions of obese people the world over. And when establishing the cause of this global problem, time and again medical experts have pointed the finger of blame at, above all else, sugar. The world's populations have long been consuming unprecedented volumes of sweeteners, with the inevitable outcome of unmanageable weight gain that is all too visible, and sufferers' health problems now tax medical services worldwide. The demand has therefore grown significantly for political action to turn the tide.

Because obesity has become so widespread so quickly there is a temptation to think of it as a uniquely modern problem, that obesity was something that earlier generations knew little about and rarely discussed. This is not the case. Overweight people were regularly discussed and portrayed in earlier periods and, as often as not, they were ridiculed. For centuries, overweight people have been traditional targets of abuse and scorn. If we want to think seriously about obesity, it is important to take a longer historical view, and we could do worse than listen to children. The young have their own way of capturing popular attitudes and moods – though sometimes in the harshest of fashions.

Through the years, larger children have traditionally been the target of cruel schoolyard humour and ridicule, and the list of nicknames aimed at overweight children is long, inventive – and often cruel. Many of us can perhaps remember such names from our own childhood. If not, we can turn to the remarkable research of Iona and Peter Opie into the playground language and play of British children in the mid-twentieth century. The Opies captured terms that are often crude and cruel, but sometimes imaginative and inventive. However, for the unfortunate victims of such barbs, they can be perceived as malicious and deeply upsetting. Who would enjoy being called, among many other things, balloon, barrel, bouncer, Falstaff, fat belly, glutton, jelly-wobble, lardy, piggy, porker, plum pudding, steamroller, Tubs . . . or, for girls, Bessy Bunter, Fatima or Tubbelina? Equally harsh terms are used for greedy children – greedy-guts, dustbin, piggy, hungry-guts – although often the greedy and overweight tend to be viewed as one and the same. Conversely, similarly hurtful terms are hurled at thin children, although the greedy and the thin tend not to attract the harsh cruelty reserved for the overweight. It also seems that children have made up their minds about larger people by the time they are six; research shows that, on the whole, they don't like them. All this is part of an astonishingly complex popular culture which thrives among children at play, in the school yard and in their free time out of school.[1] But this 'jocular hostility' towards overweight children also reflects a much deeper, almost ageless, culture of poking fun at fat people.

While it is hard to find much humour in the story of modern obesity, fat and significantly overweight people have traditionally been laughed at. British culture is littered with fat figures of fun – overweight characters who became enduring, popular figures on the cultural landscape. Most famous of all, perhaps, is Falstaff – fat, boastful, vain, gluttonous and utterly unprincipled – a figure of rotund fun around whom Shakespeare wove a humorous, but poignant, account in three of his plays. In the eighteenth and early nineteenth centuries, it was the turn of graphic satirists and caricaturists to make great play with overweight people – not least with *living* people who were significantly overweight and prominent at the royal court and in Parliament. We have Hogarth's overweight judges; Rowlandson's gluttonous diners; and the abiding images by Cruikshank and Gilray of a healthy, rotund 'John Bull' – that mythical figure of the indomitable Englishman, generally seen as a stout, assertive figure, challenging all-comers with his fierce patriotism – in contrast to an emaciated French revolutionary Jacobin. Such images of the obese flit in and out of graphic caricature throughout the period.

Dickens, too, included obese people in his writings, most memorably Mr Pickwick, as did Lewis Carroll and John Tenniel with Tweedledum and Tweedledee, their names quickly entering the vernacular to describe identical issues or people – but in the original they were obese. The tradition was continued in the twentieth century when new characters emerged as figures of fun in popular British culture, and who were distinctive because of their size or weight. We have the Reverend Awdry's 'Fat Controller' in the *Thomas the Tank Engine* book series; and Frank Richards created Billy Bunter of Greyfriars School for a hugely popular boys' comic. Bunter was a fat, greedy, obnoxious youth whose weekly appearance in *The Magnet* (and later on TV and in films) became an enduring cultural image.

He was shadowed at much the same time by another rotund character – the fat lady of the risqué seaside postcard. Best remembered in the work of Donald McGill, the round-bottomed, large-breasted lady on holiday, dominating her diminutive, hen-pecked husband, remains – to this day – a favourite item at seaside resorts.[2] And with the arrival of silent movies, and later with sound, an early favourite was the American comic Fatty Arbuckle, whose name quickly became a standard insult among schoolchildren.

Today, however, sufferers of obesity are not seen as a joke but as a serious problem. We are currently experiencing unprecedented levels and degrees of obesity, with whole swathes of the population becoming obese in their early years. Before long, some entire societies will be obese, as overweight people begin to dominate the population. At the current rate of increase, for instance, obese people may form a majority of the population of the USA and Britain by 2050. Obesity on this scale is new, and the emergence of people who are significantly overweight as an obvious and inescapable feature of modern life has evolved within living memory. What was, not long ago, unusual, has become commonplace.

Overweight children and adults, and the costly facilities of all kinds to cater for them, all this is now so common to us that it often passes unnoticed; it seems merely part and parcel of who and what we are. At the same time, obesity is creating an array of problems for society at large, the most pressing being an almost inexhaustible list of related ailments and physical infirmities. Although obesity itself is not an illness, modern medicine finds itself deeply troubled and taxed by the different illnesses which are directly driven by obesity.

Curiously, the early signs of what would become the modern epidemic of obesity first revealed themselves in Pacific islands, where supermarkets and the associated new lifestyle utterly transformed local life. But the spread of obesity in those islands was too distant, too remote from the West's beaten track to catch the eye. British medical researchers began to notice it in the 1950s, then in the '60s and '70s, when it began to take hold in South America, and then the Caribbean. Again, it tended to be overlooked, initially, because the most pressing medical efforts were directed at stamping out hunger and malnutrition. Doctors and researchers tended to ignore the local rise of obesity and its consequences because their attention was focused elsewhere. When people became fatter, this seemed at first sight a small price to pay for ending the hunger that had previously stalked those and other societies.[3]

Obesity became unavoidable under the all-determining shadow of globalisation, when so much of the world turned to fat. Urbanisation, car transport, TV and modern media, modern shopping habits – and the arrival of Western food outlets – all these served, very quickly, to transform societies into variants of their Western prototype.

Yet, at first glance, the rise of modern obesity seems mysterious. The initial response was to blame sugar – but that raised a curious problem. People on both sides of the Atlantic were buying much less sugar than their parents and grandparents, but at the very same time they were getting fatter. In the last twenty years of the twentieth century, domestic homes kept much smaller stocks of sugar in the kitchen or larder than in earlier periods, and people were adding much less sugar to their food and drink at home than in any previous period since the industrial revolution. Yet at the same time, they were getting fatter.

The explanation, obvious today but contentious and disputed over the past generation, lies in the nature of changing food and drink habits which have become so widespread. People no longer need to add sugar to their diet because it is being added for them by food and drinks manufacturers. Moreover, sugar is added in volumes which, at times, beggar belief. And as mass-produced food and drink have come to dominate people's diet, the people who consume those commodities have become fatter and fatter.

Two students of obesity have recently remarked that 'the whole world, rich and poor, young and old, is getting fatter'.[4] The evidence is all around us, and the data is astonishing. People today are bigger and heavier than at any other time in global history. The total numbers of those deemed overweight and obese around the world are bad enough; worse still, perhaps, is that out of those shocking statistics, 170 million children under eighteen are overweight or obese.[5]

The evidence is also visible – inescapable, if unscientific – wherever we care to look: overweight and obese people are a standard feature of modern life. It is most striking in the West, often shockingly so in the USA, but obesity has become a global problem and it is equally noticeable in rapidly developing societies, notably in Asia.

The question of obesity and the great variety of serious health issues it brings is now a regular cause for debate in the media. Childhood obesity, illnesses caused by obesity, the health cost of dealing with obese people, the strain on health services . . . all these and more regularly make for arresting headlines. We hear more regularly now about the difficulties of squeezing the obese into airline seats, for example, and the economic, logistical and welfare considerations that surround air travel. In Britain, in a five-year period, the NHS spent £7 million adapting equipment to cater for obese patients – bigger beds, wheelchairs and mortuary slabs. And more than 800 ambulances have been designed or altered to cater for obese patients.[6]

Designers, architects and planners have to take account of mankind's increasing size. When the new Yankee Stadium was opened in 2009, it contained 4,000 fewer seats than the previous stadium, opened in 1923, because wider seats were required to cope with the increased average size of the fans. The older seats had been 18–22in (46–56cm) wide; today they are 19–24in (48–61cm). Similar data, apparently trivial, is available from all corners of modern American life, as institutions and businesses seek to accommodate the expanding size of the American people. Ferries in Puget Sound have also increased the width of their seats; ambulances in Colorado have been fitted with winches to handle excessively heavy patients. Even undertakers have had to respond by making larger coffins to accommodate obese corpses. A standard coffin is 24in (61cm) wide, but an extra-large version of 37in (97cm) is now available.

Such evidence – unscientific, even frivolous perhaps – offers a low-level snapshot of a major US problem. But the fundamental underlying issue is simple – and serious. An estimated one in three Americans are obese, a doubling of the figure in a mere three decades.[7] More than two thirds of the current US population are overweight. Not a single state in the Union has an obesity rate below 20 per cent: twelve can claim rates of 30 per cent. And it is getting worse. By 2030, there will be a predicted 65 million more obese people living in the USA.[8]

The hard demographic evidence is available in any number of forms and is analysed most tellingly perhaps by the National Center for Health Statistics, located at the Center for Disease Control and Prevention. In the forty-two years to 2002, American men and women grew taller by an inch but, in the same period, the weight of the average American man increased from 166lb to 191lb; for women, their average weight increased from 140lb to 164lb. Similar increases were also recorded in boys and girls.[9] In 2003, obesity in the USA was roughly 32 per cent of the adult population; a mere ten years later, it had risen to 38 per cent. By 2010, it was estimated that more than 65 per cent of Americans were either overweight or obese. There are also major ethnic differences, with African-American adults recording levels of obesity that reach 48 per cent. Women fare even worse – 57 per cent of African-American women were obese in 2011–14.[10]

This is not a uniquely American problem, and the cost of coping with such widespread obesity has become critical across Western societies. Everywhere, from Scandinavia to the USA, the medical cost of caring for the obese is substantially higher than for other patients. In the USA, despite variations from state to state, the annual current cost runs to $210 billion.[11]

The seriousness with which modern medicine takes obesity can be gauged by one simple index – there has been a massive increase in the publication of specialist medical literature devoted to the subject. The words 'obese' and 'obesity' now appear regularly in medical and academic literature. Indeed, the word 'obese' appears in the title of no fewer than 19,770 articles and books published in the decade ending in August 2007. Almost 13,000 of those works appear in a mere five years between 2002 and 2007.[12]

By the early twenty-first century, the levels of obesity in the USA were causing significant alarm at the highest levels of government. No less a figure than the US Surgeon General issued a 'Call to Action to Prevent and Decrease Overweight and Obesity'. Even the US Department of Agriculture became involved, explaining how, between 1970 and 2010, the number of calories consumed by Americans had increased by 25 per cent. This is the equivalent of an extra meal each day and was a direct result of the type of food consumed. Put simply, Americans have developed unhealthy eating patterns and, although it is true that Americans are drinking less sugary soda, their consumption of sweeteners remains high because of the sugar added to their highly industrialised foodstuffs.[13]

While the USA offers some extreme examples of the modern scourge of obesity, other countries are rapidly following a similar route. This is largely because of a dramatic switch in global diets as people adapted from traditional, local and generally healthier diets to the consumption of highly processed Western food and drink. In the process, obesity has taken hold globally. Mexico worries that its children have become the fattest on earth.[14] And in middle-class communities in Delhi, according to a WHO report of 2005, 32 per cent of men, and 53 per cent of women, were thought to be obese. Indeed, one Indian in five is considered overweight. Not unrelated, an estimated 75 per cent of foreign investment in India has been in highly processed foods.[15]

China, exposed to Western foods only in the past generation, now has 350 million people who are overweight, and 60 million regarded as obese. One quarter of the Chinese population now falls into these categories. Bizarrely, it is thought that, at the same time, about 100 million Chinese are undernourished, a reminder that obesity and malnourishment can accompany each other even in the same society.[16]

France, too, has registered a sharp rise in obesity, from 5.5 per cent in 1992 to 14.5 per cent in 2009.[17] But the British top the European league table, closely followed by their Irish neighbours. In the space of little more than a single generation – thirty years – obesity has tripled, and threatens, at the current rate of increase, to have half its population recognized as obese by 2050. With very good reason, one recent publication described Britain as 'the fat man of Europe', with one Briton in four classed as obese in 2013. Medical researchers argue that Britain has already become an 'obese society', where being overweight is 'normal'.[18] The estimated cost of all this to the NHS – which bears the brunt of obesity and related illnesses – by 2050 will be an estimated £10 billion.[19] The cost is already enormous – currently £5 billion.[20]

Even more startling is how quickly this has come about. In the USA, for example, the number of overweight Americans doubled in a mere twenty-five years.[21] Today's levels of British obesity are three times the levels of 1980. Then, only 6 per cent of men and only 8 per cent of women were obese; today, 25 per cent of the British population is obese. The broad outlines of this story are well known because it is so obvious and unavoidable. Anyone under the age of thirty, perhaps, might not be so acutely aware of the problem for the simple reason that they have grown up with it. But any observant middle-aged or older Briton need only cast their mind back to their own childhood to realise how differently they studied, played, worked, travelled, dined and enjoyed themselves. As the nation became more sedentary, more inactive, more addicted to convenience food and drink, individuals have become heavier, and society is ever more beleaguered by the consequences.

Today, people walk less and drive more. In Britain, one in five journeys by car are for less than one mile. The British also spend six hours a day in sedentary pleasures – TV, computers, reading – and eat high-calorie, mass-produced foodstuffs that often contain huge quantities of sugar. This mix of inactivity and unhealthy diet has resulted in the British becoming a nation that consumes many more calories than it requires for its increasingly sedentary lifestyle.

Despite differences between men and women, and variations between ethnic groups, the overall trend is indisputable. So, too, are the consequences: obese people run the risk of catastrophic illnesses. An international study confirmed that obesity leads to type 2 diabetes, hypertension, myocardial infarction, angina, osteoarthritis, stroke, gout and gall bladder disease, colonic and ovarian cancer. It is also thought that obesity places great mechanical stress on the body and may even lead to sleep problems, breathing difficulties and back trouble. This cluster of health issues for overweight people is known as 'metabolic syndrome'.[22] And all this is in addition to the problems of social stigma, low self-esteem and an overall poor quality of life. In Britain alone, it is estimated that 30,000 deaths each year are caused by obesity, and 18 million days of sickness and absence from work are directly attributable to obesity and its adverse effects.

When parents of obese children are asked for an explanation for their children's size, they readily point to their offsprings' lifestyle, and especially to the amount of time watching TV, or using a tablet, laptop or computer.[23] It is there that they are exposed, for hours on end, to cleverly devised adverts promoting food and drink which is nutritionally worthless, but which is often rich in sugar.

The problem is now so severe and widespread that health services and a coalition of medical experts regularly urge government to act. They are not only desperate now to persuade the population at large to adopt a healthier lifestyle, but also to influence that powerful lobby of commercial interests – 'Big Food' – whose products have encouraged the stampede towards unhealthy choices over food and drink. The aim of such critics is to reduce the volumes of sugar, fats and salt currently saturating the mass-produced, processed foods and drinks which contribute so significantly to national obesity.[24]

Though few doubt the evidence of the rapid increase in obesity, the precise causes remain contentious, even among medical and scientific experts working in the field. There are even those who choose to see the entire issue as yet another 'moral panic', one of those periodic social alarms which have gripped societies over the centuries. Some sociologists have been keen to tease apart the social origins of different waves of mass anxiety in very different historical and social settings. Where once it was witches, communism, muggings, football hooliganism, mods and rockers or AIDS, some have now turned their attention to those who are grossly overweight. And although it is true that the debate about obesity has generated a huge and growing scientific literature, much of it is dogged by disagreement, and by the special pleading of vested interests.[25] For all that, the core, demographic evidence is irrefutable. Time and again, doctors and medical sociologists point to simple but persuasive data of the rise in numbers of obese patients.

The most troubling aspect of the entire story is the extent of childhood obesity. The warning bells first sounded in the USA. In the twenty years to 1995, the number of overweight children increased from 15 to 30 per cent. A decade later, researchers thought that the problem had spun 'out of control' in Europe. Indeed, it had increased twice as fast in England as it had in the USA. But most other European countries – Poland, Spain, Italy, Albania and Greece – were following close behind. Even France, fiercely protective of its cuisine and associated lifestyle, had begun to succumb. Similar data began to emerge from Asia. In Japan, childhood obesity doubled between 1974 and 1994. In Thailand, it increased by 3 per cent in three years between 1990 and 1993. Even in Saudi Arabia, 16 per cent of boys aged 6–18 were found to be obese in 1996.[26]

What linked these very diverse geographical locations was a curious but critical fact. Childhood obesity rose fastest, and remains most tenaciously rooted, among low-income groups. It has settled into a universal law: those who are most susceptible to developing obesity tend to be those with low incomes. 'Only the very poorest inhabitants of the poorest countries offer an exception to the fate formula . . .' These are the very people who lack the money for (and even the access to) fresh fruit and vegetables; they are 'struggling households' which 'stock up on sugar, starch, oil and other processed foods – high energy and low costs'. One study puts the matter bluntly: 'Slimness is becoming an unattainable luxury for the poorer families in our midst.'[27]

This had become very striking in Britain. By the early years of the twenty-first century, the data was astonishing. In 2011, three in ten British boys and girls aged 2–15 were overweight or obese. Astonishingly, perhaps, between 2011 and 2013, sixty-two children under eighteen underwent weight-loss surgery. There had been only one such case in 2000.[28] Although children's overall health improved dramatically in the course of the twentieth century, by the twenty-first century childhood obesity presented a startling setback. It was recorded in all fifty US states, among boys and girls – although it was most prominent among African-Americans and American Indians. Again, the hospital costs of caring for obese children and youths were excessive and had risen from $35 million in 1979–81 to $127 million in 1997–99.[29] Moreover, obesity continues to rise among American children; between 2006 and 2008, it rose from 15 to 20 per cent among 6–11 year olds. Critics pointed their finger at sugar. The American Heart Association became so alarmed in 2009 that it issued a recommended level for sugar: 'High intakes of dietary sugar in the setting of a worldwide pandemic of obesity and cardiovascular disease have heightened concerns about the adverse effects of excessive consumption of sugars.' The recommended limits – five teaspoons of sugar for sedentary women, nine for men – sit uncomfortably with the actual intake of twenty-two teaspoons. These suggestions prompted a broadside of commercial and of sponsored scientific denunciation from all corners of the US food industry. Sugar had become so central to 'Big Food' – in effect it had become the lifeblood of a massive, multi-million-dollar industry – that it was not about to be staunched by the reasonable but ineffectual pronouncements of the medical lobby.[30]

A report for the WHO found that childhood obesity was rising the world over; it stood at perhaps 2–3 per cent of all children aged 5–17. It was highest in the Americas (30–35 per cent) and in Europe (about 20 per cent). In sub-Saharan Africa, it was a mere 1 per cent. 'In most countries, there has been documentation of a rapid increase in the prevalence of obesity among children.' Between 1980 and 2000, it rose sharply in Australia, Brazil, Canada, China, Spain, the UK and the USA. The report concluded that, among children, 'as in adults, over-weight and obesity are common and becoming increasingly common in populations throughout the world'.[31]

In the autumn of 2016, the World Obesity Federation painted a very gloomy picture indeed of a contagion of child-hood obesity worldwide. Pacific islands – Kiribati, Samoa and Micronesia – had easily the worst statistics in terms of propor-tion of populations, but they were closely followed by Egypt with 35 per cent of people under the age of seventeen obese or overweight. There followed, in order, Greece (31 per cent), Saudi Arabia (30 per cent), the USA (29 per cent), Mexico (29 per cent), and the UK (28 per cent), followed closely by France and the Netherlands. Not surprisingly, then, an estimated 3.5 million children worldwide have type 2 diabetes. Many, many more have ailments which are directly linked to obesity.

Viewed globally, there has been a 60 per cent rise in childhood obesity since 1990, and the patterns once thought peculiar to the West are now replicated worldwide. In a mere decade, the proportion of children who are obese or overweight have increased from one in ten, to one in eight. In Britain, obesity is growing twice as fast among children as among adults. It is thought that one third of all Europe's obese children are British. But the problem is far worse in the USA: an estimated 32 per cent of American children were thought to be overweight or obese in 2009–2010. Among British children aged 2–11 years, 14 per cent were found to be obese in 2004. For 11–15-year-olds, it rose to 25 per cent. By the early years of the new century, 'obesity is now the most common disorder of childhood and adolescence'.[32] The problem, again, is diet. Most of the victims are from low- or middle-income groups. Wherever researchers looked, the pattern of causation was the same – fast food and carbonated drinks, and even a global decline in breastfeeding in favour of baby formula milk. Today, one half of the world's inhabitants live in urban areas and most children do not get adequate exercise. And everywhere, they favour sweet, fizzy drinks (the sales of which have increased by one third in the past ten years), and fast food from Western-style outlets. Sugar is everywhere. In Egypt, it is heaped in tea five or six times a day.[33]

The likely medical consequences for obese children are well documented: psychological ill health (bullying seems to be a common hazard); heart trouble; breathing difficulties; inflammation; diabetes; orthopaedic problems; and liver disease. In addition, childhood obesity not only brings its own health problems, but lays the foundations for obesity in adult life. People do not 'grow out' of obesity; obese children are highly likely to become obese adults – with all the related ailments. Each of these illnesses may have their own medical solution, but the attack on the root cause – obesity – is not so much a medical as a social question.

The core problem is diet. We also know that eating preferences endure for a lifetime, and this simple fact is critical and basic to the activities of food and drink manufacturers, and the advertisers who promote their products among the young. Advertisers and food manufacturers know that if they can capture a child's loyalty – if they can cultivate a child's taste and commitment to their products – they will have them for life.[34]

One of the early consequences of children's eating habits – of consuming sweet foods, notably breakfast cereals – is the early onset of dental problems. Among British children, for instance, the story of dental health provides an extraordinary example of the impact of the modern diet, and especially of the role of sugar in that diet. British medical authorities have become increasingly alarmed about children's poor dental health. In fact, concerns about dental issues first emerged more than a century ago, but today they illustrate much wider anxieties about children's diet, and especially about the levels of sugar in that diet. Even in the earliest days of dentistry in the nineteenth century, British dentists regularly complained about the levels of tooth decay and poor oral hygiene among the nation's young. The problems came into sharp focus with the late-century establishment of compulsory schooling and the obligatory medical inspection of all children in schools. Medical examinations confirmed what many had long suspected – the existence of a wide range of health problems, notably, of course, among the poor. Time and again, doctors and dentists recorded high levels of poor dental health. The extent of ill health and of poor health facilities (plus the cost of medical care) in time became a powerful political impulse behind the determination to improve the nation's health by the introduction of a free national health service. Despite the post-war NHS, and despite the subsequent seventy years of free medicine, serious dental problems continue to plague large numbers of British children.

In 2005, a specialist dental division of the Royal College of Surgeons stated it was 'seriously concerned about the state of children's oral health in England'. The reason was straightforward – almost one third of five-year-olds suffered from tooth decay, and the most common reason why 5–9-year-olds were currently admitted to hospital was because of dental problems – sometimes 'for multiple tooth extraction under anaesthetic . . .'[35]

Notwithstanding the shortcomings in dental provision (some areas have no appropriate dental care for the young), the underlying problem is diet and parenting. Many parents remain astonishingly unaware of the need to encourage their children to adopt appropriate dental healthcare. Routine teeth cleaning, regular visits to the dentist, and attention paid to a child's diet – as obvious as they may seem, these basic actions need to be encouraged. In the words of another report, 'Parents and children should be educated about the risks of tooth decay and the importance of good oral health and prevention.' Successful campaigns in Scotland and Wales had led the way, and England needs to follow suit. Local authorities were also urged to implement fluoride schemes where they are lacking. Most significant, perhaps, the report urged that 'efforts should be made to raise awareness of the impact on tooth decay and explore ways to reduce sugar consumption'.

Behind all this lay some stark and indisputable evidence. Of the 46,500 children and young people admitted to hospital for multiple extractions, the largest numbers were aged 5–9, and dental decay was the most common reason for surgery on children in that age group. Despite three generations of national health cover, and in one of the world's richest countries, children's dental health was poor.

The problem was, however, uneven, with striking regional variations. Not surprisingly, the nation's poorer regions showed the highest levels of poor dental health among the young.[36] The north-west of England (home to a string of struggling industries and the towns that sustained them) had much worse dental problems among children than the prosperous regions of the south-east of England. But everywhere, the basic point is simple, and bluntly stated by dental experts: 'It is lamentable that tens of thousands of children need to be admitted to hospital when poor oral health is largely preventable. The cost of all this is an estimated £30 million.'[37]

Quite apart from the pain involved, children's dental problems create other significant difficulties. They result in eating and sleeping problems, children missing school, and parents having to take time off work to take children to emergency dental appointments. Although it is accepted that oral health has improved since the 1970s – partly because of a number of educational drives and the impact of widespread fluoride programmes – one third of English five-year-olds continues to suffer from dental problems.[38]

So far, all this is at the level of dental treatment. More important still, however, is how to tackle the problem at source; how to eradicate the cause of poor dental health? How can dental decay among children be avoided, and thus avert painful and costly medical intervention? Medical opinion is once again unambiguous: regular tooth brushing; early examination by a dentist; and, critically, parents insisting on 'a healthy diet and limiting consumption of sugar or acidic food and drink to mealtimes. Even fruit juice is acidic and high in sugar, so parents should try to give young children only water or milk.' Parents also need to be careful about medicines given to small children. 'Use only sugar-free medicines if possible.'[39]

The historical irony here is striking. For 2,000 years, from the world of classical antiquity to the present day, parents have coaxed sick children to accept unpleasant-tasting medicines by adding honey or sugar to the medicine. Now, the received wisdom is that sweetness itself is a health issue.

Throughout all this medical discussion, it was abundantly clear that English children consume much more sugar than recommended by the medical authorities. Between the ages of four and ten, English children consumed 48lb (22kg) each year, the equivalent of 5,500 sugar lumps. The recommended maximum is 8kg – and even that is regarded as excessive by some.[40] The danger is even greater for American children. The twenty-seven teaspoons of sugar they consumed in 1970 had risen to thirty-two teaspoons in 1996 – all of it added to their food and drink before they were even purchased.[41] The major obstacle is the nature of modern drinks and food, and knowing if, or how much, sugar is used as an ingredient. The history of sugar (and other sweeteners), as we have seen, has traditionally been that of an additive. People simply chose to add sugar to their food and drink according to their taste. Today, however, most of the sugar consumed is added (and often in astonishing volumes) even before the meal or drink comes to the table. The end result is that sugars are to be found 'in almost all food and are the most important factor in the determination of oral health. It is especially problematic in children who have become accustomed to sugar at an early age.'[42]

The list of food and drinks which contain sugar is broadly familiar, and most of them are promoted by clever and lavish advertising campaigns. That list has also grown ever longer with the industrialisation of foodstuffs and the complex process of refrigeration and chilling, to say nothing of the chemical experimentation on the various ingredients used. The obvious foods include sweets and chocolates, cakes and biscuits, fruit pies and puddings, breakfast cereals, jams and honey, ice cream, fruits in syrups, sweet sauces and ketchups. Apart from the famous carbonated, canned sodas, a range of other drinks have sugar added: fruit juices, cordials, sports drinks, caffeinated energy drinks and yoghurt drinks.[43] Many of these items pass into children's hands as snacks, as refreshments, as pacifiers – or when children themselves simply demand them. It was no surprise to hear the Director of Public Health England in 2015 stress 'the need to urgently reduce the amount of sugary snacks and drinks in our children's diets'.[44] The main obstacle, however, is the massive power of a food industry that has come to consider sugar as a vital ingredient – especially in children's food, and the millions of dollars and pounds spent on targeting children.

British children have long enjoyed a sugary diet (the poor, as we saw, traditionally relied on jam in the late nineteenth century), but the problem has become more acute and more damaging in recent years. Modern British children are thought to consume 'thirty times the amount of soft drinks and twenty-five times the amount of confectionery they did in 1950'. The volumes of soft drinks consumed doubled between 1992 and 2004. It was as if post-war austerity gave way to a cornucopia of sugary drinks and food for children. These sweet temptations were dangled before children by a new and powerful advertising industry which had itself been transformed into a multi-million-pound business, especially via the explosion of

TV and, in Britain, after the advent of commercial TV from the mid-1950s. Thereafter, TV became as much a means of advertising as it was of entertainment. By the early twenty-first century, the British food industry was spending £450 million a year on advertising – three quarters of which was directed at children. In 2001–02, for example, Coca-Cola spent £23 million, Walkers Crisps £16.5 million, and Müller's pot dessert £13.15 million. Food advertising fell into four distinct categories: sugary cereals, confectionery, soft drinks and snacks (mainly crisps). On children's TV, more than half of all advertisements were for food and drink. Critically, 99 per cent of that amount is to promote 'junk food'.

On both sides of the Atlantic, a veritable tidal wave of subliminal 'educational propaganda' emerged, directly aimed at children.[45] In the late twentieth century, Western parents were confronted by a problem never faced by earlier generations – an uphill battle to resist children's demands for the sweet temptations dangled before them on their TV screens.

As huge as these British figures relating to food and drink advertising are, they are dwarfed by the data for the USA, even allowing for an American population five times larger than Britain's. American children and adolescents form a major target for advertisers and their products – they watch one hour of adverts for every five hours of television viewing. In the course of a year, an American child will see 40,000 TV adverts – 80 per cent of which fall into four categories: toys, cereals, candy and fast foods.

It was in the very years that TV viewing boomed that obesity rates among American children tripled. In the late 1970s, about 5 per cent of US children were overweight or obese. By the early twenty-first century, that had risen to 35 per cent (boys) and 32 per cent (girls). While all critics accept that this was brought about by a multitude of factors, they broadly agree that advertising for food and drink played a critical role. Only the automobile industry spent more money on advertising than the US food industry – understandably, perhaps, when we realise that 12.5 per cent of all American consumer spending goes on food. More revealing still, the advertising industry specifically targets children and teenagers because they form a large and lucrative market. American teenagers currently spend $140 billion of their own money, and advertisers and their sponsors are keen to tap into that enormous spending power. It's no surprise, then, that adverts overwhelmingly targeted at the young account for remarkable amounts of advertising money: $792 million on breakfast cereals; $549 million on soft drinks; and $330 million on snacks. All this – a treasure trove of dollars passing back and forth for foodstuffs – is in addition to 'product placement', with logos and signs for certain foods and drinks carefully located on toys, in videos and movies, on the Internet and in sports stadiums. More blatant still – and even less defensible – is the placing of the same drinks and foods in dispensing machines inside American schools. In the slightly opaque words of one team of critics, it is hard to deny 'the existence of a relationship between media use and diet-related outcomes (namely, overweight and obesity . . .)'.[46]

TV has given millions of children an utterly new and almost irresistible commercial power. Dubbed 'pester power' by critics, children have been drilled (by clever marketing) to demand foods, drinks and 'treats' which serve little purpose except to quieten the babble of children's noise, and swell the coffers of advertisers and food and drink manufacturers. The irony is that the great bulk of children's foodstuffs advertised in this way is both sweet and of little nutritional value.

A 2013 study of 577 food advertisements aimed at children revealed that 'nearly three-quarters of them were promoting foods of "low nutritional quality" . . .' By the late twentieth century, such foods came in an astonishing variety of shapes, colours, textures – many ready-made for 'dunking' – and most of them laden with sugar. Some breakfast cereals manufactured and aimed specifically at children consist of 50 per cent refined sugar.[47] And it may surprise some that many of these products are apparently endorsed by an in-house, sponsored, scientific expert (dentist, doctor or researcher), whose name and approval is appended to the bottle, package or carton. It is a curious giveaway. Why should manufacturers need such support and approval, unless they face doubtful or sceptical consumers? Why even need to claim that the food or drink is healthy?

The background to all this is, of course, the serious food and health crises of the past generation – some of them catastrophic. But all of them stemming from endemic flaws in the way food is produced by the modern agri-industry. The British disaster of BSE ('mad cow disease') from the late 1980s; a crisis about salmonella in eggs in 1988; the British foot-and-mouth epidemic of 2007; dioxin found in Belgian sheep; and more recently, horse meat masquerading as beef . . . the list is depressingly long, and has had a corrosive effect on consumers' confidence in the food they bought or were served. Yet equally troubling was the related transformation in diet brought about by the food-processing industry. A range of scientists toiled long and hard in the service of major food and drink manufacturers to create tastes, flavours and sensory feelings in a wide range of drinks and foodstuffs. Their efforts have been incorporated into (and, in some cases, totally displaced) the basic foodstuffs we buy. And sugar and other sweeteners have been at the heart of this entire process. Colours that are merely chemical creations, flavours conjured up in the laboratory, tastes that originate solely from chemical experiments – all this and more is the story of processed food and drink. And hovering over the entire issue is the contentious question of genetically modified crops.

The end result has been the evolution of a distorted human diet. Nutritionists are generally agreed that a balanced diet should consist of 50 per cent carbohydrate, about 15 per cent protein, and no more than 35 per cent in fat. But the revolution wrought by food processing has significantly affected these ideal proportions, with the modern diet for many being more likely to consist of 45 per cent carbohydrate and 40 per cent fat. Moreover, the carbohydrates we consume tend to consist not of starch and fibres, 'but of sucrose, fructose and glucose – monosaccharides – or simple sugars'. The sugars we consume – spooned into our beverages, dissolved into fizzy drinks, and baked into confectioneries, sweets and treats all provide up to 20 per cent of our total energy intake. But they form 'empty calories', and are devoid of minerals, vitamins and other ingredients present in non-processed foods.[48]

This entire phenomenon has been promoted by an advertising blitz of unprecedented ferocity and ingenuity. And the barrage of commercial propaganda promoting sugary foods that are nutritionally dubious reminds one of another product that shares a similar story – tobacco. Defenders of sugary drinks and foods were quick to trot out the defence of 'personal responsibility' in matters of consumption. Should it not be the individual's right to choose what to buy or not? The promotion of sugar-laden food and drink has begun to look remarkably like the activities of the old tobacco lobby. They, too, had fallen back on arguments about individual choice. Now it was the turn of sugar and food – shouldn't parents be free to choose what their children eat and drink? And shouldn't they be able to resist their children's shrill demands for sweet breakfasts, juices and snacks? Such questions form a cynical deception – a fraudulent pretence to mask cleverly devised promotions of goods which do little but fatten and rot. The old tobacco lobby must recognize the familiar strategic manoeuvring when they see and read protestations issued by defenders of sugar and the food industry.

* * *

When the data about obesity began to make a serious impression in the latter part of the twentieth century, it prompted growing concern about sugar which evolved into a major assault on the use of sugar in food and drink. It was an assault which also targeted all the major agencies thought to be responsible for the global rise of obesity. But it prompted an inevitable counter-offensive by the food and drink industry and its lobbyists. In the USA, where that industry wields enormous commercial and political influence, lobbyists were hugely successful in rallying legislation to head off any damaging legal attacks. They knew, from earlier experience, that punitive legal cases might be heading their way. So they launched a pre-emptive strike.

In 2004–05, the US House of Representatives approved a bill with a curious but revealing title: 'The Personal Responsibility in Food Consumption Act'. The 'Cheeseburger Bill', as it became known, was designed to absolve the fast-food industry of all blame for people who became overweight via their diet, to shield the food industry 'from being sued by obese consumers'. What lay behind this bill was the recent experience of the tobacco industry. The food and drink industry was fearful of the fate of the tobacco industry which had been hit in 2004 by the threat of astronomical damages – totalling $280 billion – for the health damage caused by tobacco. Though the award was later struck down by a superior court, the lesson was not lost on the coalition of interests which lay behind the fattening diet of millions of Americans.

Despite the tobacco industry ultimately dodging culpability, by 2004 a crucial link had been established in the USA – obesity, and especially the role sugar played in the process, was being widely compared to the impact of tobacco. Whatever short-term victories of the sugar-food lobby, and however obstructionist their political allies in Congress, the tide had begun to turn against sugar.[49]

In large measure, the tide had begun to turn because attention had shifted away from the USA. As long as obesity was considered to be essentially an American problem, it seemed both contained and isolated. Clearly, no one can any longer doubt the depth and extent of American obesity. But that had initially served to deflect attention from the global problem. By the early twenty-first century, however, it had become abundantly clear that obesity was not solely an American – nor even a Western – problem. The human desire for sweetness, and the power of sugar producers, food and drinks manufacturers, and advertising agencies to satisfy that craving were all combining to produce a health crisis of global proportions. But what else was responsible for driving growing numbers of the world's population towards the dangers of obesity?

14

The Way We Eat Now

I N THE YEARS when people were buying less sugar for use in the home, the food and drink which came to dominate the dining table (or on the tray in front of the TV) were laden with hidden amounts of sugar. Anyone keen to avoid sugar in their food had to study the contents printed on the packaging. Moreover, that information was only made available by manufacturers after a prolonged rearguard battle against campaigners who demanded to know exactly what ingredients, and in what proportions, go into the foods we eat.

Today, there is scarcely a mass-produced item that arrives on the supermarket shelves without the addition of sugar. Foods which we would never associate with sweetness are prepared with a range of additives, and most notably with sugar. The presence of sugar in industrialised foodstuffs arose from two major factors. First, the long-term history of mankind's taste for sugar; and second, and more immediately, the more recent development of scientific analysis of food and human taste.

Although sugar has been embedded in the human diet for centuries, that relationship changed – quickly and unpredictably – in the late twentieth century, and it did so on the back of modern nutritional science working in league with new, high-powered food industries and their relentless promoters in the modern media.

Nutritionists working for food manufacturers discovered, in exact scientific detail, what had been apparent for centuries – that people like sugar. By the late twentieth century, science could prove what mothers cuddling a sick child at any point over the past 2,000 years could have told them – infants like the taste of sweetness. From their early days, in sickness and robust health, babies liked sugar and honey. They reacted badly, however, to bitter, sour or salty tastes.[1] Exactly why that should be so has been an abiding interest of modern scientists, their curiosity spurred (and often financed) by major food and drink companies. From the early twentieth century, physiologists had suggested that particular biological features of humans were receptive to sugar. In the 1960s and '70s, a string of research projects confirmed in humans and animals the power of sugar and sugary foods in developing cravings and excessive eating habits. Rats and human seemed to enjoy eating sweet foodstuffs; when unrestrained, they all became obese. By the end of the century, research into the body's physiological reaction to sweetness had begun to yield findings that were of great interest to the food industry. Indeed, much of that research had been sponsored by it.

The more research emerged, and the more scientists discussed the topic, the clearer it became that taste – flavour – had little to do with nutrition. People chose their food and drink because of what they expected the taste to be. And of all the tastes humans enjoyed the most, sugar seemed to reign supreme. Once this simple point had been established scientifically, it was as if the food industry had discovered the key to alchemy – how to convert something worthless into something apparently of great value. What mattered now was to create a product – a food or drink – that targeted those physiological reactions.[2]

Scientists in a range of disciplines had begun to track down the precise way sugar is absorbed into the body, and exactly how the body reacts to the presence of sugar. What emerged was the sense that humans are hard-wired to enjoy sugar. This seemed to be the case not merely with refined sugars, but also with refined starches, which the body then converts to sugar. In this way, the starch in a pizza, for instance, becomes sugar – and the brain reacts to the pleasure accordingly.

In commercial terms, however, it was research into children's reaction to sugar – the search to find what became known as the 'bliss point' in various children's foods – which opened up a treasure chest for food manufacturers. Sugar prompted pleasure in children; it also provided the energy needed for growing children, and it conveyed a 'feel-good' factor, not unlike an analgesic.[3] Major food corporations set out to devise sweet foods for children that would hit these various targets; food that would capture a child's desire for sweetness, energy and a generally good feeling. Everything seemed to revolve around the problem of providing exactly the right mix of sweetness in a particular food and drink. But the base point was simple – sugar provided the key to commercial success in food and drink.

It was, then, no surprise that food manufacturers began to target children with sweet foods. But the increasingly sweet diet of the young faced growing criticism from medical interests – not least when it seemed to be adding to childhood obesity and creating dental problems. All this was reinforced by the growing awareness that sweet drinks, in particular, were a major contributor to obesity. Equally worrying, children who were allowed to drink sweetened, carbonated drinks had come to expect *all* their drinks to be sweet. Fruit juices, sports drinks, flavoured waters – all were expected to be sweet to gain children's approval, and so attract their spending. In the very years that obesity, especially childhood obesity, emerged as a major Western problem, sweet drinks seemed to lie at the heart of the matter.

Worried observers felt sure that sugar was involved. But the exact formula – the precise role of sugar in the rise of obesity – remained unclear. This allowed defenders of sugar and sugary diet to dispute the criticisms of sugar as a major contribution to global obesity.

From the mid-1960s, the fight was on. In the sugar corner, the producers of sugar itself and the food manufacturers who poured sugar, in vast and growing volumes, into their products. In the opposite corner, health professionals, medical staff and a growing band of activists determined to alert all and sundry to the dangers of the sugary concoctions that had come to form the bedrock of the modern diet.

* * *

The American food industry in the 1950s was driven forward by the concept of convenience, which resulted in an ever-changing array of easily managed and quickly cooked products. The word 'cooked' was hardly appropriate – heated, boiled or grilled became the dominant forms of preparation, and all that was made possible by new kitchen gadgets. Women – the whole phenomenon was aimed at women initially – who remained loyal to old-fashioned food preparation found themselves surrounded by younger women who found the labour-saving food products easier and more convenient. Despite the hard work of those school teachers who stuck to the skills of domestic science, American domestic cuisine swiftly succumbed to the culture of convenience.

It is easy to see why. For the growing number of women who worked, these changes in diet and cooking came as an indisputable blessing. Domestic chores – notably cooking for a family at the end of a hard day's work – were eased aside by the availability of foods that needed no time-consuming preparation. Once again, the food industry saw the challenge, and inaugurated its own domestic science lessons conducted by company employees who were eager to promote the virtues of convenience products for the hard-pressed American housewife. The industry invented imaginary cooks who promoted the new foods, replied to fan letters and figured in promotional literature – but who never actually existed. Fantasy home-makers endorsed invented foods for a nation that had become seduced by the television screen. Millions now dined from a tray as they watched the evening television programmes, all of them punctuated with copious adverts, many of which were for food and drink. In the process, the teaching of old-fashioned home economics effectively died out, aided by subversive propaganda from the food industry. *Time* magazine caught the mood and the times perfectly when a 1959 article on convenience food carried the headline, 'JUST HEAT AND SERVE'.[4]

Along with the ease of serving, consumers of the new convenience foods were also absorbing unacknowledged volumes of sugar; few of the new convenience foods were produced without the addition of plenty of sugar. All this, pioneered in the USA, very quickly spread to other corners of the world, hastened by the rapid evolution of new, global food corporations. Many of those US-based corporations gobbled up local companies in distant parts of the world and transformed them into producers of foodstuffs and drinks perfected in the USA. People round the world began to follow the USA not merely in popular culture – movies, TV and music – but in the food they ate. And what they ate was highly processed – and sugar-rich.

The story of industrial foods began at the American breakfast table in the 1880s, with the efforts of Dr Kellogg, fresh from medical work in New York, who aimed to transform America's breakfast. His small-scale experiments with wheat products in his home kitchen and a small medical facility at Battle Creek, Michigan, laid the foundations for the company that bears his name. For centuries, gold prospecting had been a significant activity in the region; and when Dr Kellogg added sugar to his cereals, it was as if he, too, had struck gold.

Rivals followed with similar products – also laced with sugar – and the commercial war began for dominance of the American breakfast table. The food companies all used the new tactics and ploys of advertising to promote their products. These were the great days of the billboard, with colourful adverts plastered on sides of buildings, hoardings, on buses and trams, in newspapers and magazines – indeed, on any available space that might catch the eye of the paying public. Advertising – and heaps of sugar – transformed breakfast cereals into a basic ingredient in American daily life.

Despite in-house concerns in the 1940s about the amount of sugar added to new products, the combination of wheat cereals laced with sugar proved an astonishing commercial bonanza. By the late twentieth century, 85 per cent of the US breakfast cereal market was dominated by three giant companies, all guided by marketing and advertising men who scrutinised regular, detailed social trends to find the best way to generate greater expansion and sales. By the last quarter of the twentieth century, the key to further commercial success was the critical fact that young mothers were going out to work. In the early morning rush to prepare and feed children before dashing to work, convenience at the breakfast table become the pressing concern. They also needed to capture the imagination of the children themselves.[5]

The story of sweet products aimed at children began effectively in 1949 when Post Foods found itself struggling to sell its breakfast cereals against the competition. Up to then, cereals had been heavily promoted as a healthy alternative to the traditional fatty breakfast of bacon, sausages and spam. Post Foods now began to sprinkle their cereals with sugar and children loved what they were offered. Competitors in the food industry followed suit with a string of sugary inventions aimed at the American breakfast table. Thereafter, breakfast became a battleground between food giants and their advertising agencies. All sides began to add sugar in abundance until, eventually, 50 per cent of all American breakfast cereals contained added sugar.

Scientists in the food laboratories and marketing men on Madison Avenue now tinkered with a host of innovations to increase the appeal of breakfast cereals to children. Cereals began to appear in different shapes and forms (as letters, for example) and appeared on the supermarket shelves, and increasingly on TV screens, under a host of new, catchy names. In the decade from the mid-1950s, dazzled by extravagantly budgeted advertising campaigns for food and drink, and led primarily by General Foods, America was 'led to a different way of thinking about food'.

The men at the helm of these massive companies grasped a simple but essential point. The foods they produced and sold 'had to be easy to buy, store, open, prepare and eat'.[6] America had entered the world of convenience food, some of which were so far removed from their original, natural state, they were unrecognizable – food and drink concocted by scientists in conversation with marketing men, tested by social scientists, and with the commercial results analysed by the most up-to-date mathematical formulae. Through all this, sugar was rarely far away. The invention of breakfast drinks that required only the addition of water – and sugar to taste – proved a massive hit, both for the manufacturers and the sugar industry. So, too, did 'pop-ups' – a type of cake or pastry pocket with a filling that was heated in the toaster. They came, like the drinks, in a variety of flavours – and all with ample additions of sugar.

The consequent sales of breakfast cereals were astonishing – $660 million worth in 1970 rising to $4.4 billion in the mid-1980s. Efforts to curb the power – the stranglehold – of the three biggest companies failed in the teeth of their dogged legal and political defence. Like the great trust battles of the late nineteenth century, the companies could wear down their opponents by the immense wealth and influence they were able to throw into the battle to defend their interests. No less troubling was the companies' refusal when pressed to disclose to their customers just how much sugar was being added to their cereals.

At the same time, it was becoming clear that America's recently acquired diet was creating major health problems, most notably in dental health. One American dentist, concerned at the poor dental health he encountered among his young patients, acted on his own initiative. He bought seventy-eight brands of cereal and examined their content in his own laboratory. One third had sugar levels between 10 and 25 per cent, another third up to 50 per cent, with eleven even higher. There also seemed to be a correlation between the sweetest cereals available and their aggressive TV advertising directed at children.[7]

In addition to concerned parents, a host of critics emerged challenging the cereal manufacturers to explain and reduce their production of heavily sweetened cereals. The companies fought back, of course, sometimes changing the name of products, sometimes dropping items, and rethinking how they promoted their foods. By the late 1970s, however, they were also under scrutiny from federal agencies, themselves galvanized by dental professionals who provided damning evidence about the extent of dental ill health among the young. These pressures, at a time of rising consumer power, were led by energetic activists not easily cowed by commercial giants, and yielded results. But flawed legislation in Washington that was easily by-passed, and the enormous financial and commercial power of the food lobby and advertisers, ensured that proposed curbs on advertising aimed at children foundered.

All was not lost, though. Data about sugar and children was becoming more readily available. It was now undeniable that American children were being targeted by TV adverts about sugary foods. The food manufacturers and their agents knew that the adverts worked, and they were relentless in promoting sugary cereals. By the late 1980s, however, the companies had begun to remove the word 'sugar' from their promotions, and even from the name of the cereals. Even they had finally realised that sugar had been categorised as 'bad'; it had become a word that conjured up a string of unhealthy qualities, a commodity which parents should try to avoid when feeding their children.[8]

Throughout the 1990s, the major food companies were engaged in a broad commercial battle against consumer groups agitating against sugar, against new competitors in the field and then – breaking ranks – against each other in a price war. Throughout the entire story, their in-house scientists and marketing men were devising new ways of presenting cereals; creating snacks and products which could be represented as cereals in another guise – in the form of a 'handy' bar, for example. Yet, time and again, new products were blended with volumes of sugar and launched at the children's market. Sometimes, the adverts for such products were not only misleading but downright untrue, although criticism was often deflected by the companies' reliance on slowing down to a bare minimum the handling of these assertions, and the time-absorbing complexities of legal processes.

Activists worried about the impact of sugar on America's health had come to realise that drawing attention to foodstuffs was only part of their task. They also needed to challenge the major drinks corporations about their excessive use of sugar in the drinks that millions of Americans turned to throughout their waking hours. Soda drinks had become much more American than apple pie, and those drinks came bursting with sweetness. So it was, in the last decades of the twentieth century, that alarm grew in the USA about the consequences of the nation's rapidly changing diet – Americans were getting fatter, and large numbers of children were suffering unusually high levels of dental decay at a very young age. Any scrutiny of the immensely powerful food and drink industries was bound to invite a prolonged and brutal conflict. Moreover, those companies were no longer simple American or British entities, but had morphed into mammoth transnational corporations of unprecedented wealth and influence.

Today's global food systems are controlled by a small group of massive corporations that possess an unprecedented concentration of economic power and money. A small number of commercial behemoths dominate the food market at every level – the market for essential agricultural products, the manufacture of food, and food retailing. Oxfam calculated in 2013 that 70 per cent of the world's food systems were controlled by no more than 500 companies which are the prime users of all the commodities which go to make up our industrial food and drink. Within that small band there is a tiny elite of mega-corporations, the corporate giants who control much of what we eat and drink. Among them are household names, most notably Nestlé, Unilever, Mars, Coca-Cola, PepsiCo, Mondalez, Danone, Associated British Foods, Kraft, General Mills, Kellogg's, McDonald's and Compass Company. The power of such organisations almost defies analysis. Nestlé, for example, enjoyed a revenue in 2012 that was larger than the GDP of all but seventy of the world's nations. Their sales figures that year, of $100 billion, compares to Uganda's GDP of $51 billion. And the emergence of this agri-industrial power play is not merely a commercial process, but has been encouraged and, indeed, aided by Western governments, primarily via massive subsidies in euros and dollars.

The origins of this global food system began after the Second World War, and the desperate need to restore a shattered European continent. Between 1947 and 1952, the US Marshall Plan pumped $13 billion into Europe, much of it in the form of American food, animal feed and fertilizer shipped into Europe. It was a programme which pulled Europe back from the brink, revitalised the continent, and also helped catapult the USA to global dominance by 1950. After 1945, governments across western Europe were equally resolved never to repeat the errors of the past, and were determined, among other things, to enhance European food production. One priority of European governments, coalescing into what became the European Community, was the creation of large-scale, healthy agriculture, assisted by subsidies. The eventual outcome was that European food became plentiful and affordable, and farmers were generally well rewarded for their labours.

Fifty years later, however, these subsidised agricultural schemes were creating massive surpluses – the 'food mountains' and 'wine lakes' that were so often mentioned in the media. Much of those surpluses ended up on the world market where they had the effect of undermining producers in poor countries. When the World Trade Organisation was launched in 1995, it was designed, among other things, to end such subsidies and remove trade restrictions. What happened, however, was that poor countries were forced to open their markets – notably under pressure from the World Bank and the International Monetary Fund (IMF) – while richer countries maintained their subsidies for local agriculture.

Predictably, poor countries were overwhelmed by European and US produce. Moreover, that produce had been greatly helped by subsidies. Recent analysis of subsidies has confirmed that in both the USA and in Europe, the bulk of such subsidies go to major corporations, a process that has been dubbed 'corporate welfare'. Moreover, the largest recipients of European subsidies, as reported in 2009 (from evidence unearthed by the Freedom of Information Act) were the major transnational sugar companies. Sugar, along with other crops produced by the major corporations, was being procured at guaranteed prices for a specific period of time.[9] By this method, cheap, subsidised sugar passed into the diet of untold millions. In the USA itself, sugar had long been a highly protected and subsidised industry. *The Wall Street Journal* claimed, in 2015, that the absurdity 'of the federal sugar program is legendary'. The US tax system saw to it that sugar producers were guaranteed profits, whatever the market conditions.[10]

By the early twenty-first century, however, the era of such subsidies was drawing to a close. In many respects, it was too late – the damage had been done. Millions of people had already been weaned on food and drink that was harmful to their health. The greatest obstacle to changing the system, however, is the global power of multinational corporations, which effectively control the world's food supplies. Which single nation state – including the USA – can bring to heel corporations who owe no loyalty to any particular state, and who wield unprecedented power? They are organisations who can simply move their money and their plant elsewhere, and adapt their systems using cheaper labour in alternative locations. What had happened to the world's food is that it had become global – a particular but central example of the broader phenomenon of globalisation – owing no obligation to any government or people.

Who can resist the decisions and blandishments of such fabulous wealth and power, or tell the owners and managers of such corporate fortunes what they should or should not do? Moreover, for these companies, the arithmetic is simple – foodstuffs now deemed unhealthy are highly profitable, whereas wholesome, healthy foods are much less so. Wholesome foods yield profits of 3–6 per cent; highly processed foods 15 per cent.[11] Which corporation Chair or Board is about to propose reversing the order to its shareholders? The end result is a global problem. In the words of the UN's Special Rapporteur on food, 'Our food systems are making people sick.'[12] They are also making them increasingly fatter.

The underlying factor behind this extraordinary problem was the nature of the food and drink disgorged by the major food industries. What people eat and drink had been utterly transformed in little more than fifty years, by corporations that controlled agriculture, food processing and even food retailing. And at the heart of these global changes in food and drink lay the question of sugar.

The origins of the European processed-food industries can be traced back to the 1860s.[13] Before then, food and drink traders were largely locally and regionally based. All began to change with the advance of modern industry. By the 1870s, for example, the availability of artificial ice made possible the long-distance transport of fresh fish, although from the 1920s fish was also cheaply available in tins; thereafter, tinned salmon became a household favourite. New metal steam-powered mills were introduced to grind American wheat which was harder than European varieties and had been imported from the vast farmlands of North America.

By 1945, all this had fallen into the hands of large-scale industrial enterprises. Even the bread made from that wheat was increasingly produced in factories. Evaporated milk, with large amounts of sugar added, became another popular item in poorer homes because of its taste and the fact that it was long-lasting. Yet the diet of poor people remained rooted in the old staples – all made palatable by the addition of sugar, which itself now passed through modernized, dockside refineries. In the form of jams and syrups, sugar continued to prove its value to low-income groups across Europe.

Such changes were a mere prelude to the intensive industrialisation of food after the Second World War. This was largely a technological revolution, brought about by scientists and nutritional research conducted both by the major food industries in their own laboratories and in university research labs. As food and drink became ever more industrialised and scientifically rooted, the food markets themselves changed – large-scale markets, swift transport systems and storage, all became integral features of the new food industries.

The end results were sometimes hard to recognize as food – 'pink slime' and 'meat slurry' – both meat products – being perhaps the most repugnant. Yet both were sold to the public as 'hamburgers'. 'Turkey Twizzlers' and other similarly bizarre inventions were the extreme offspring of the industrialisation of food. Indeed, a large number of meat products had the actual meat content reduced by processes of separation, spinning, boiling and freezing, before being bulked out with water, flavourings and colourings. Here, again, sugar was important as an additive.[14] Even today's ingredients lists on lower-value packages of ham, turkey, hamburgers, sausage and other cold meats make interesting reading.

* * *

This extraordinary upheaval in global dietary habits was not an accidental by-product. It was cleverly devised and executed by companies that had learned how to exploit human needs and responses – and to play to them via a scientifically and commercially inspired use of sugar as a critical ingredient.

We need only examine the daily cycle of meals and eating to sense the impact of sugar on our lives. Even before vast numbers of people step from their homes in the morning, their sugar intake has begun its impact on their bodies. The day often begins with breakfast cereals, many of which arrive at the table with a heavy dose of sugar, or sugar is added to the bowlful. Toast, muffins and other breads – or pancakes, in the USA – contain their own helping of sugar. Even the morning fruit juice may have been sweetened for them. Tea or coffee is often taken with sugar to taste.

There might be a number of mid-morning, sugar-laden biscuits, pastries, cakes or snacks to keep energy levels – and enjoyment – at their peak, and then lunchtime continues a similar pattern. Sugar is not only consumed via the obvious sweet foodstuffs – fast foods, above all, and yogurts, for example – but even in foods which suggest a healthier option, such as salad dressing, for instance. Sugar is even present in savouries that, again, might not seem sugary when tasted – sauces with pasta, ketchups, bacon, processed meat products, hams and other cured meats, with the accompanying rolls and bread. Unless we ask for tap water, even the bottled drinks we might choose to drink during the day contain their own additives in the form of syrups, sugar, fruit concentrates or synthetic sweeteners. Anyone choosing a dessert is bound to be offered artificially sweetened foods in a variety of forms, and these foods tend to be much sweeter in the USA than in Europe. An evening meal, at home or dining out, is similarly likely to be infused with the same sweet additives, especially if that meal is a takeaway dish, or pre-cooked, chilled or frozen and is simply reheated in a microwave.

Clearly, large numbers of people do *not* eat like this; but hundreds of millions do. Moreover, even people who seek to avoid food with added sugar have to make a special effort when shopping; they need to be vigilant about what they purchase, and carefully read labels on bottles, packages and bags. Even so, reading the label is sometimes not enough to avoid sugar, because the contents often come in a format incomprehensible to anyone but a chemist. It was no accident that the food industry put up a rearguard fight against the demands to oblige them to describe in intimate detail the content of the foods they manufactured. They knew that what they would reveal might act as a warning bell to more careful shoppers.

Sugar, then, is a regular companion to processed foods, but it operates in a very different fashion from sugar in natural foodstuffs. While a banana might contain 16g of sugar, a chocolate bar might contain 40g. And when that sugar is combined with fat as part of the process, it has an altogether different physiological impact. Furthermore, because processed foods are stripped of protein and fibre, which would normally absorb sugar slowly, a bigger 'sugar spike' is caused.

All this may seem, even to the curious, an accidental by-product of modernised industries: manufacturers of food and drink creating highly efficient manufacturing techniques designed to bring cheap food speedily to the shopper in the supermarkets. It is true that employees of the food industries scour the globe looking for suitable products for their customers, but they are supported in their search for new foodstuffs, back home, by teams of scientists working at the micro level of food and drink. Traditionally, their basic concerns have been with sugar, fat and salt – the basic ingredients of much that constitutes the modern diet.[15] Each has been studied by scientists and mathematicians to explore the best ways of incorporating those ingredients into a particular food and drink. Sugar has, for example, been reduced in the laboratory to a simple fructose additive that will boost the allure of the food to which it is added.[16] At its most extreme, food scientists have not so much tinkered with foods as *invented* new foods – some of which owe their very existence to experiments in laboratories. Moreover, the more successful foods have been sweetened specifically to enhance their appeal to children.

Food manufacturers have invested untold millions of dollars, pounds and euros changing, improving and inventing food-stuffs and drinks 'that are quite literally irresistible'. In 1985, for instance, General Foods had a research budget of $113 million.[17] What they were searching for were products that would hit what became known as the 'bliss point': 'For all ingredients in food and drink, there is an optimum concentration at which the sensory pleasure is maximal. This optimum level is called the bliss point.'[18]

The concept of a 'bliss point' was coined in the 1970s by a Hungarian mathematician, and was quickly adopted by the food industry as a means of promoting their products.[19] In the course of the 1990s, what had been a vague but interesting concept solidified into a hard and fast scientific fact – a belief that had a certain plausibility was now accepted as a scientific formula and a commercial device.

At various late-twentieth-century gatherings of food scientists, food executives and advertisers, the concept of the 'bliss point' settled into a key concept in their technical vernacular. Here was a word – bliss – that they could use with some abandon, knowing that it conjured up happiness. Moreover, it was a form of happiness that seemed to be grounded in empirically proven fact. Time and again, the food industry was urged not to worry about the suggestion of happiness associated with the use of the word 'bliss'. After all, when they sold their food and drink, they were promoting tastes that people enjoyed. Nutrition was a minor issue which, in the minds of most customers, did not even register when they scanned their supermarket aisles for their daily sustenance.

Above all else, what offered the 'bliss point' more than any other single commodity was sugar. The reason is straightforward; in the words of one commentator, 'Humans like sweetness . . .' The trick is getting the quantities just right. The aim of food and drinks companies was then to ensure they consistently hit the 'bliss point' for sweetness.[20]

When food and drink executives convened at their international gatherings, this and similar messages was music to their ears. It was delivered to them by various market and scientific researchers. Bliss – pleasure – was quantifiable and could be doled out by the sugary spoonful and added to an endless range of foods and drinks. Moreover, it was confirmed by an abundance of research into the physiology of taste – studies of how the tongue's receptors transmitted sensations to the brain, and how the brain responded to the different tastes entering the body. That research also suggested that the brain demands still more pleasure when sweetness enters the system. The individual is prompted to eat even more of the initially sweet delight. All this – a complex science which lies on the frontiers of scientific and neurological research – was naturally of great interest to the food industry. Their new target was to harness what science had told them to the best commercial advantage – to persuade people, via their complex nutritional chemistry, to like a product so much that their body demands more of the same. It is, in fact, the pattern of behaviour that shapes a variety of addictions. In this case, it is a craving for sweetness – and then still more sweetness.[21] The appreciation of the science of addiction led the food industry to add critical additives to their products – and none more important and successful than sugar. Sugar and sweeteners became, in effect, the bait around which they could promote a product. People loved a sweet taste, and demanded more of the same on a regular basis.

If the scientific and nutritional researches of the late twentieth century into taste and sweetness provided lucrative opportunities for the food and drink industry, the very same evidence was equally valuable to their opponents, more especially those who began to point to the links with obesity. While sugar was just one of a string of food and drink additives that fell under the scrutiny of scientific and governmental bodies worried about the wider problems of nutrition and well-being, there was mounting evidence to suggest it played an even more corrosive role than its early critics had imagined.

The initial campaign against sugar was American, and arose from the concern about the role of sugar in processed and instant foods, and the links to obesity. Growing numbers of parents were concerned about what they learned about the chemistry of new foods – fears about the artificial flavours and colourings, and the huge volumes of salt, fat and sugar added to those foods. It coincided with worries, reported by parents themselves, about the hyperactivity of their children. Was it caused by the very food the children *liked*?

Again, the food industry rallied its forces – its scientists, lobbyists and marketeers – to discredit criticism and reassure customers of the healthy nature of their products. Their arguments were rooted in the broad acceptance that babies and the young liked sweetness from their first days; a love of sugar was a natural, physiological urge and not to be denounced or dismissed by critics. The food industry was merely providing – so they claimed – in modern forms and varieties, the food and drink which humans naturally loved and craved. We now know, however, that behind the science of the food industry there lay some intricate political machinations, and downright deception, to hide what they knew to be the realities behind the impact of sugar.

When the chorus of criticism against sugar began to grow in the 1960s, the sugar and food lobby adopted a number of different tactics to head off its opponents. We now know, from very recent research, that they took steps to *buy* supportive scientific support. Over a period of fifty years, sugar interests have subsidised research that would deflect and minimise the truth about the impact of sugar on obesity. They did this by turning the scientific spotlight on to other ingredients. With substantial funding from the sugar lobby, a team of Harvard scientists reviewed scientific research in such a way as to minimise the importance of sugar, and push the blame for obesity on to fat. Fat, not sugar, was firmly placed centre-stage in the rising debate about obesity. This discovery – in the papers of deceased Harvard professors, and only exposed in September 2016 – raised a much broader and more sensitive issue, namely the role of industrial sponsorship in scientific research.

The Harvard revelations prompted wider concerns about the way scientific research is often sponsored by major corporations. It became apparent that a raft of researchers were being funded by those with a vested interest in sugar. And a similar story unfolded in Britain, of scientists pursuing their research courtesy of funding via the sugar lobby. This ought not to be so surprising, though. Such co-operation is long-standing and widely accepted in many areas of corporate and political life – necessary even. But the whole issue left a string of very real concerns.

It emerged that *all* the major food and drinks industries used scientific studies to deflect criticism of their unhealthy ingredients, and particularly sugar. After a fashion, that might not be thought unusual. The food industry was, after all, a vast, complex concern which had for decades used scientific research to devise and improve its products. What was revealed in 2016, however, was substantially different. Sugar executives had set out, from the mid-1960s, to devise research, and to encourage findings, that would remove the spotlight from sugar. To do that, they needed amenable researchers, in prestigious institutions, willing to help – in return for financial support.

The revelations of 2015–16 exposed what some had long suspected – that the food industry was specifically paying scientists to produce reports favourable to their products and their general interests. The long-term consequences of that research, launched in 1967, served to divert attention *away* from sugar towards other possible causes of obesity. It proved to be a highly successful tactic and, for the best part of half a century, sugar was effectively exonerated.[22] In the process, it also led to the denigration and belittling of serious researchers who wrote about the dangers of the excessive use of sugar in our modern diet.[23]

At one level, this was just the latest twist in a remarkably old and durable story – of sugar's hold over US politics and strategy. However, by 2016, it was equally clear that matters of public health had become of prime concern – and not merely in the USA. Sugar's influential grip could no longer be allowed to exist unchallenged. It was clearly playing a corrosive role in the widespread decline of the nation's health and well-being. What made the task more daunting was that sugar lay at the heart of many enjoyable features of modern life – not least, the relatively recent and rapidly spreading social habit of dining out.

One major change in eating habits in the last fifty years has been the growing interest in visiting restaurants. Until recently, eating outside the home was, for millions, a very special treat. Today, it is an unremarkable event. Between 1980 and 2000, one half of the budget spent by Americans on food went on dining out. Equally striking, Americans ate substantially *more* when dining out. In the years since 1950, portion sizes of food served to customers in American restaurants increased four-fold.[24] Meals outside the home are eaten at work, at school, in a canteen, or food is bought to the workplace from local food outlets. Institutional food (provided, say, by schools and the workplace) has to be cheap and plentiful, and that invariably means processed foodstuffs with sugar as a significant additive. This presence of sugar in processed food is much more striking in fast-food restaurants and takeaways. A burger or fried chicken with chips and a cola – followed by a frozen dessert – are all heavily laden with animal fats, with sugar being a predominant carbohydrate. Typically, such a meal can easily amount to 1,600 calories.[25]

Yet such fast food is now simply inescapable – outlets dot the high street, shopping centres and are signposted along major roads and highways. The food they offer has come to dominate the diet (and hence the health) of tens of millions of people. The USA, again, led the way, and eventually became the most seriously affected by the consequences. By 2001, the USA was home to more than 13,000 McDonald's outlets, 5,000 Burger Kings and more than 7,000 Pizza Huts. In the twenty-five years to 1995, the number of fast-food meals eaten by Americans increased fourfold. In Britain, the number of McDonald's restaurants doubled in the decade to 1993, while in Europe they quadrupled between 1991 and 2001, from 1,342 to 5,792, and there was a similar growth of Burger King and Pizza Hut. More astonishing still, the rate of growth was even faster in Asia. By 2004, Indians and Chinese were eating fast food more frequently than Americans.[26]

As if this were not startling enough, the fast-food revolution is paralleled by changes in the way people eat at home. Pre-cooked, chilled or prepared dishes (main courses, vegetables and desserts) have taken the place of meals that are prepared and cooked in the family kitchen. We have already seen the reasons for this – convenient, cheap food has offered an easy alternative to the more time-consuming effort of creating nutritious meals from scratch. In millions of homes – even when those homes are equipped with a state-of-the-art kitchen – the key appliance is now the microwave. From 1970 to 2000, the microwave came to dominate the Western kitchen. In that time, more than 90 per cent of American and Australian kitchens possessed a microwave. The only exception to this growing trend was France, a society which holds on to its more traditional cuisine and dietary habits. The increase in the use of microwaves was also less significant in developing countries where refrigerators were a priority – notably, of course, in hot countries.

Often, the family meal is eaten while watching TV, which has come to dominate breakfast-time as well, when children, particularly, are watching. This habit was, inevitably, of great interest to food manufacturers. Market researchers told the food manufacturers that traditional meal-preparation was changing very quickly (in parallel with the rising proportion of women in the labour force) and the food companies were quick to step in by devising and marketing a string of ready-made, pre-cooked or chilled meals that could be easily prepared for a couple or family – without requiring any cooking skills at all.[27]

The food industry set about 'methodically to move all forms of cooking from the kitchen to the factory . . .' They were helped, again, by science and by the development of new plastics with which to package, cover, bottle and seal their products. The end result is now familiar – all that is required is for the 'cook' to remove the vacuum-sealed package and pop the entire meal into a microwave oven for the requisite number of minutes printed on the label. What could be simpler?

These major changes in the way we eat, and *what* we eat, have become global habits almost imperceptibly. They are widespread throughout the West, but they have also become even more striking, and with more fundamental changes, in the developing world. The West took a long time to emerge from societies where hunger (and even starvation) was commonplace, to the current state of widespread obesity. Developing countries have undergone the same transition very recently, and at breakneck speed. Once haunted by malnutrition and undernourishment, many are now plagued by obesity. Yet it is now clear that obesity has become a form of malnutrition.

What reinforced the changes brought about by the global drift towards industrialised food and drink has been a trend which, again, tends to be overlooked because it is so basic to the way we now live. What we eat and drink has been refashioned by the way we shop.

Shopping, for the modern consumer, has been transformed and, indirectly, this transformation in shopping has itself become a factor in the rise of obesity. New shopping patterns have, in their turn, created major consequences for *what* people eat and drink. The most important change – obvious, but generally unnoticed – has been the rise of the modern supermarket, which has affected not only *where* people shop, but what sort of food is available to them. Since there are now an estimated 20 million supermarkets in operation worldwide, they have come to exercise a massive influence over vast numbers of people. As supermarkets proliferated, millions of people got fatter. But what is the connection?

The origins of the supermarket can be traced to the late nineteenth century and the development of the new department stores and chains in Europe and North America. In the first half of the twentieth century, consumers began to move gradually towards buying industrially produced foodstuffs, but it was after 1945 that the process speeded up. Just as the industrialisation of basic foods was viewed in Europe as creeping 'Americanisation', so, too, did the arrival of the supermarket herald a revolution first pioneered in the USA. In Britain, the 175 supermarkets in 1958 had risen to 2,110 by 1972. By then, Germany had 2,802 and France 2,060. France, though deeply resistant to changes in eating and drinking habits, nonetheless succumbed and, by the late 1980s, 56 per cent of the French food market was in the hands of supermarkets.[28]

The rise of the supermarkets inevitably resulted in a precipitous fall in the number of independent retailers. In the last quarter of the twentieth century, the UK lost 120,000, West Germany 115,000, France 105,000, and Spain 34,000. All lost out to the modern supermarket. Today in the UK, the top five supermarket chains control 70 per cent of grocery sales. In the USA, the five largest control 48 per cent. In the Netherlands, ten supermarket chains control 75 per cent of the market. In the years when food and drink were being manufactured by ever fewer, ever larger conglomerates, food outlets were also declining in great numbers. The outcome was a convergence of massive food manufacturers and food outlets, and they were making hugely influential decisions about what and how their customers should choose and prepare their food.[29]

Supermarkets introduced a totally new way of shopping. They removed the shop counter, and made their food available 'for the taking'. They invite shoppers to take what they want. But what those customers *see*, on the packed shelves before them, is carefully presented and strategically placed. Moreover, the massive supermarket chains can also dictate to the producers – to farmers and manufacturers – precisely what they want; they dictate the size, shape, colour, volume – and price – of food and drink on their shelves. Thus supermarkets have come to shape the consumption of food, and much of that consumption has increasingly involved packaged and ready-made cooked foods – most of it with copious amounts of added sweeteners.[30]

* * *

Since approximately 1945, food itself has been utterly transformed by the rise of 'agribusiness' – major corporations creating massive agricultural businesses along the lines of corporate models – and the emergence of what critics have termed 'Big Food'. In the USA, for example, half of all the nation's food is produced by a mere ten corporations. Much of that food is processed, and most of it with the accompaniment of sugar. And it is also worth bearing in mind that one nation's processing of food is never restricted to domestic markets. The West exports enormous amounts of processed foodstuffs to other nations. Britain, for example, exports £19 billion worth of food annually. But of that, £11 billion is highly processed, and £6.4 billion only lightly processed. Only £1.4 billion is unprocessed.[31]

Sucrose lies at the heart of most of these processed foods and drinks. It is true that a range of sweeteners are now used instead of, or as well as, traditional cane sugar, but sugar retains its popularity as an ingredient. In the words of one study of sucrose, 'The nexus of sugar (sucrose) and confectionary products is inescapable ... Nothing yet devised by humans or nature has the unique sweetening, bulking and manufacturing properties of natural sugar.'[32] And as we have seen, sugar pervades the manufacture of almost every type of food imaginable – dough and sponges, biscuits, icings and fillings. It appears in dairy products, ice creams, custards, frozen desserts and yoghurts. Processed foods use sugar 'to import body and texture [and to] produce bulk ...' Tinned fruits and vegetables often contain sugar, as do ketchups and chilli sauces, as well as pie fillings, desserts, cured meats, bacon and sausages.[33]

But it delivers its most extravagant hit in breakfast cereals. About 30 per cent of these cereals arrive at the breakfast table pre-sweetened, with the most sugary of cereals consisting of more than 50 per cent sugar. Breakfast has become *the* main meal for sugar. And all this is in addition to the sugar in soft drinks, in alcohol and fruit preserves, jellies and jams.

Behind all these foodstuffs, there lurks a simple point – sugar is ubiquitous. It resides in many of the foods we consume, and food manufacturers regard it as an essential ingredient which adds flavour, bulk and texture to an enormous variety of consumer products – foods, drinks, cosmetics, medicines . . . and so the list goes on. It seems, on its own and in combination with other ingredients, to lurk behind the world's growing problem of obesity. And if any single type of food or drink was to embody this problem of sugar-driven obesity, and perhaps even exacerbate it more than any other, then we need look no further than sweetened soft drinks.

15

Hard Truth About Soft Drinks

THE EARLY SUCCESS of sugar as an additive to hot drinks was as nothing compared to the impact of sugar consumed in soft drinks – carbonated or still – in the late twentieth century. Indeed, the global consumption of sugar (and later of other sweeteners) was utterly revolutionised after the Second World War by the emergence of the soft-drink industry. The range of non-alcoholic, water-based, still or carbonated drinks is now enormous, and most of them have been developed as highly sweetened – until a very recent campaign against them in the UK and the USA. Often flavoured by a variety of fruits, berries and sometimes even by vegetables, most were highly calorific – so calorific, in fact, that they have become both a major cause of obesity, a topic of fierce political debate and even a matter of punitive taxation. How did this happen to drinks that had their origins in a simple refreshment on hot summer days in the American South?

Non-alcoholic soft drinks – cordials and home-made fruit drinks – have a long history as both a medicine but mainly as a thirst-quencher. Although popular in the West from the sixteenth century onwards, they belonged to a much older tradition reaching back to classical Rome, and drinking spa waters. Suspicion of polluted water also encouraged experiments in creating aerated mineral waters, most famously and enduringly by Jacob Schweppe in 1792. At London's Great Exhibition in 1851, for example, 1 million bottles of aerated water were sold. Such drinks were promoted for their medicinal qualities, but they really took off commercially as simple refreshments. When new flavours were added to them – ginger was the most popular – they became hugely popular on both sides of the Atlantic.

In the USA, local pharmacists created their own soft drinks, and competition developed to market new versions – new flavours, sweet and fizzy – for the growing urban population of North America and Europe. Sarsaparilla, root beer, dandelion and burdock, these and many others competed for customers' cash. Eventually, many of these drinks ditched their therapeutic claims and simply offered refreshment. They were also warmly supported by the temperance movement as an alternative to the evils of alcohol, although their popularity was much more broadly based than that. There were millions of people who wanted, and liked, refreshing, cooling, cheap drinks – especially in hot American summers – without any of the disadvantages of alcohol. At the heart of all these drinks lay one ingredient which was to dominate the business from the late nineteenth century to the early twenty-first – sugar.[1]

Cheap soft drinks proliferated throughout the Western world in the late nineteenth century: Rose's Lime Juice (lime juice had its own long history in naval use); concentrated fruit drinks (from Australia); barley waters; orange squash; blackcurrant juice; cranberry juice; and, in the USA, grape juice. The soft-drink revolution in the USA really stemmed from the 'soda fountain' invented by a Yale scientist, Benjamin Silliman, who had a Yale college named after him, and who sold soda water by the glass and by the bottle. Like many others, the drink was first promoted as a medicinal drink. But when flavours were added, 'a whole new industry was born'.[2] New, more elaborate soda machines were devised which allowed customers to choose their favourite flavour, and 'soda parlours' quickly established themselves as an indispensable feature of American urban life. By 1895, there were an estimated 50,000 of them in the USA. When, in that same decade, safe new bottles were invented, soda could be sold in a bottle, and could be bought as a take-away item for consumption at home, in restaurants or in parks.

A large number of new flavours were devised by blending sugar with fruits, vegetables, herbs and flavourings, and American customers could choose from an astonishing variety of sweet fizzy drinks. Enterprising manufacturers were constantly in search of new brands of drinks, while vigilantly keeping their recipes secret from competitors. At first, most of them claimed their drinks possessed medicinal qualities. Both Dr Pepper and Coca-Cola were launched on the American market in the 1880s and were heralded with medicinal claims, but these were soon forgotten in the wave of commercial success they enjoyed at the nation's soda fountains. When the two companies franchised their secret syrups to bottlers, 'the modern soft-drink industry was born'.[3] There followed a quite astonishing story.

Although US soft-drinks manufacturers offered a wide range

of flavours and ingredients, it was drinks with the addition of caffeine from the kola nut – and later, equivalent substitute flavourings – which rapidly outstripped all others and, by 1920, 'cola' drinks dominated the US market.[4] In 1930, more than 7,000 US bottling plants were turning out 6 billion bottles annually, and Americans drank it in enormous quantities. In 1889, they drank an estimated 227 million, 8-fluid-ounce drinks; by 1970, that had risen to 72 billion. In the process, a curious cultural image evolved – sweet fizzy drinks became the very representation of America itself, and the best-known brand names seem to capture the essence of modern America. Coca-Cola (1886), Pepsi-Cola (1898) and Dr Pepper (1885) had all originated in the US South, and all began life in local pharmacies where 'soda jerks' experimented with their own sweet, fizzy concoctions. It was as if they had struck liquid gold. The American per capita consumption of their drinks grew by leaps and bounds, from 0.6 gallons in 1889, to 3.3 gallons in 1929, 23.4 gallons in 1969 and 44.5 gallons in 1985 – and ever upwards thereafter. Today, an increasing proportion of sales now takes place outside the USA.[5]

The most famous, and most globally recognized of all these drinks is, of course, Coca-Cola, and the modern-day data for that product makes for amazing reading. In 2012, the company sold its products in more than 200 countries, with 1.8 billion daily servings of the drink – that is one drink for every four people on the planet. Coca-Cola is the twenty-second most profitable company in the USA; its revenues exceed $48 billion, and its net income is $9 billion. 'By the twenty-first century, Coke had conquered the globe, its market reach unmatched.' All this from a commodity which had been a patent medicine in 1886.

Remarkably, the company was able to keep the price of a bottle of Coke at five cents from 1886 to 1950. Their vending machines, scattered carefully along America's network of highways and bus stations, all dispensed bottles in return for a five-cent coin. Like most of its competitors, Coca-Cola was a very sweet product. The original formula mixed 5lb of sugar into every gallon of syrup. By 1900, every 6fl oz serving of the drink contained four teaspoons of sugar. The result, even as early as 1910, was that Coca-Cola 'was the largest industrial consumer of sugar in the world'. By then, some 100 million pounds of sugar were consumed annually via Coca-Cola.[6]

The key element in the success of Coca-Cola was cheap sugar, which, in the early twentieth century, was in plentiful supply from tropical producers and from US beet farmers. Moreover, the entire system was aided by the US Government's policy of subsidies and tariffs for sugar. The price of sugar had fallen at the precise time the new soft-drinks companies began to pour unprecedented volumes of sugar into their new inventions, and the appetite of those producers for sugar seemed insatiable. Coca-Cola, for example, used 44,000lb of sugar in 1890. Thirty years later, that had grown to 100 million pounds. The US Government's support for the sugar industry in effect bolstered not only the prosperity of the sugar corporations but also the fledgling drinks companies that relied on cheap sugar. It is no surprise then that the US political and strategic interest in sugar-growing regions became a key element in US foreign policy. It was as if the interests of US sugar and US foreign policy worked as one.[7]

Coca-Cola continued to thrive, even during the First World War when, along with other companies, the company had to reduce its sugar consumption by 50 per cent. Even here, though, the company was able to turn things to their advantage, via publicity which stressed its devotion to national duty by following the new restrictions. The post-war fall in sugar prices was a boon to the soft-drinks companies, although Pepsi was temporarily bankrupted by poor investments in the Caribbean. At the same time, the soft-drinks industry had become a major lobbyist in Washington and was keen to maintain the supply of cheap sugar.

Throughout the 1920s and 1930s, the soft-drinks companies were key lobbyists on all matters to do with US sugar policy and, in the same period, one company – Coca-Cola – established itself as *the* dominant soft-drinks manufacturer. The company took off through a string of clever commercial decisions and via persuasive advertising, as well as the help offered by federal health experts' approval for their main drink. Their success was also linked to the massive interwar expansion of US road transport. Coca-Cola located new dispensing machines at 500,000 petrol stations, while travellers by bus were served by machines in the nation's bus terminals. The company also began to expand its overseas operations, developing outlets in twenty-eight countries, from Burma to South Africa.[8]

The volumes of sugar involved in this expansion were huge. At around the outbreak of the Second World War, Coca-Cola alone devoured 200 million pounds of sugar annually, and clearly needed protection for its vital sugar supplies – at low cost. When war returned, the US Federal Government once again stepped in to stabilise sugar prices, although the drinks companies baulked at sugar rationing, as they had during the First World War. Coca-Cola returned to its tactics of the earlier war, of persuading the public – and especially the Government – that Coca-Cola was a patriotic company, and that its drinks provided a vital wartime boost; it was a reviver and refresher for hard-pressed, war-time workers, and was especially important for men in uniform. In what proved to be the transformation of the product and the company, Coca-Cola became a wartime necessity, not a trivial luxury.

Alongside cigarettes, Coca-Cola was heavily promoted as vital for the war effort, both at home and abroad. Specially commissioned reports and adverts promoted the same idea. Even the US Surgeon General was recruited to the task:

> In this time of stress and strain, Americans turn to their sparkling beverage as the British of all classes turn to their cup of tea and the Brazilians to their coffee. From that moment of relaxation, they go back to their task cheered and strengthened with no aftermath of gastric repentance.

The master stroke, however – a move of incalculable commercial value – was securing military backing for Coca-Cola. The US Army persuaded the Government to exempt Coca-Cola from sugar rationing and to allow the drink to be sent to US bases across the USA, and to all theatres of war. In January 1942, General Eisenhower ordered monthly supplies of Coca-Cola for the US military. Thus, the company bought its sugar at a Government-pegged level, then had exclusive access to the vast market that was the US at war – in Europe and Asia. Company profits soared to $25 million in 1944 alone.[9] It was a serious blow to Pepsi-Cola (only recently recovered from bankruptcy), in that it was not allowed access to the wartime military machine. Coca-Cola's privileged position enabled it to race far ahead of all its commercial rivals. In the course of the Second World War, Coca-Cola sold an estimated 10 billion bottles at military bases and supply (PX) stores, providing 95 per cent of all the soft drinks sold to the US military.[10]

This global reach was attained by the company piggy-backing their wares on the vast US military machine. After Pearl Harbor, Robert Woodruff, chairman of the company, had patriotically declared, 'We will see that every man in uniform gets a bottle of Coca-Cola for five cents, wherever he is and whatever it costs our company.'

The unquenchable thirst of US servicemen for Coca-Cola spread the taste to all corners of the globe. Staff from the Coca-Cola company (nicknamed 'the Coca-Cola Colonels') travelled in the wake of the military, setting up bottling plants and distribution systems to reach the troops. No less important, senior US military figures – Paton, MacArthur, Omar Bradley and, above all, the Supreme Allied Commander in Europe, Eisenhower – effectively endorsed Coca-Cola. Eisenhower and General Marshall signed orders authorising the shipping and installation of Coke plants from the USA to the European theatre – and all this at a time of scarce shipping capacity for vital military equipment.[11]

From the start of the US involvement in the Second World War, senior military figures had appreciated that Coca-Cola was good for morale. Eisenhower was the most prominent figure who believed that the drink kept his men happy and fighting fit, and his commanders echoed that belief, signing requisitions for Coke from one corner of the world to another. More surprising, perhaps, the military actually paid for the transportation and erection of Coke-making machinery in all theatres of war, and the labour behind all this activity was military-based. Although the company dispatched 248 of its employees to supervise those operations, it was army and navy engineers who provided the muscle and skills. By the end of the war, the US military had constructed sixty-four bottling plants for Coca-Cola, many of them employing GIs as workers. The impact of all this was sensational. Between 1941 and 1945, the US military bought 10 billion bottles of soft drinks from Coca-Cola.

No one doubted for a second that Coca-Cola was vital. For the millions of Americans in uniform, it was a reminder of home. In the USA itself, Coke's advertising told Americans that the drink was patriotic. It was the American way of life in a bottle. That same image became global and people the world over came to regard Coca-Cola as quintessentially American: 'a symbol of our way of living'.[12] Throughout the wartime advertising campaign, the company traded relentlessly on this theme, in print, on radio and via a host of educational publications. Above all, however, it was the homesick young American serviceman – in New Guinea or North Africa – who most valued Coca-Cola. Extracts from the many thousands of letters they sent home are testament to that:

. . . *I always thought it was a wonderful drink, but on an island where few white men have ever set foot, it is a Godsend* . . .

. . . *But the other day, three of us guys walked ten miles to buy a case of Coca-Cola, then carried it back. You will never know how good it tasted* . . .

. . . *The crowning touch to your Christmas packages was the bottled Coca-Cola. How did you ever think of sending them!*

. . . *To have this drink is just like having home brought nearer to you* . . .

. . . If anyone were to ask us what we are fighting for, we think half of us would answer, the right to buy Coca-Cola again . . .[13]

The phrase, 'Coca-Cola', was even coined as a military password when the Allies crossed the Rhine into Germany. Nor was the power of that drink lost on the Axis powers, although they used it to denigrate the USA. In the words of one Nazi propagandist, 'America never contributed anything to world civilization but chewing gum and Coca-Cola.' In fact, the war confirmed that when Coca-Cola passed into the hands of friend and foe as a gift from an American GI, they liked what they tasted. Here was the seedbed for a remarkable commercial phenomenon – the global spread of post-war taste and demand for Coca-Cola. From Africa to Fiji, from India to Iceland, locals got their first taste of Coca-Cola from the American military in wartime.

* * *

Alongside this military-assisted commercial coup, the Coca-Cola company appealed aggressively to influential and moneyed people around the world, inviting their investment in new plants for Coca-Cola, arguing that the drink was good for local business and development. Magnates from India to Brazil secured contracts to bottle and distribute Coca-Cola. The company failed, however, to persuade the US Government that the company should be part of the Marshall Plan for Europe's reconstruction, and despite Coca-Cola's immense wartime global reach, its immediate post-war fortunes flagged. Pepsi-Cola, on the other hand, revived, courtesy of dynamic new management and aggressive advertising. But this changed, once again, in the 1960s under fresh management, and with the massive expansion of Coca-Cola outside the USA; more than forty new plants opened in 1960 alone.[14]

What transformed Coca-Cola within the USA was the post-war rise of a new dining phenomenon: McDonald's and the other fast-food chains – Taco Bell (1946), Burger King (1954) and Kentucky Fried Chicken (1952). All of them proliferated along the highways constructed under President Eisenhower's stewardship. Following Coca-Cola's earlier model, McDonald's franchised their restaurants and, by 1960, there were 250; in 1970, 3,000. The founder, Ray Kroc, was also a huge fan of Coca-Cola and made that company the exclusive supplier of soft drinks to McDonald's expanding chain. By 2000, McDonald's had become the largest customer for Coca-Cola. It was a perfect deal for the soft-drinks giant – throughout this expansion, they could satisfy the enormous appetite for their drinks with very little investment in infrastructure, because bottling and distribution was licensed out to local companies. As we have seen before, there are few who have come close to creating alchemy – but now Coca-Cola were as close commercially as it is possible to be, transforming a humble soft drink into unimaginable, golden profits.

The man who led Coca-Cola to its pre-eminence in the post-war world was Robert Woodruff, whose wartime success was the introduction of Coke to all corners of the world on the back of the US military. Like the manufacturers of breakfast cereals, Woodruff also appreciated the importance of securing the loyalty and the taste of the young. American children were introduced to Coca-Cola in their early years – at those formative moments when habits developed in the heart of the family, with playmates and the local neighbourhood – and those habits would endure for ever. Coca-Cola's post-war triumph was to establish consumer loyalty on a scale that other companies could only marvel at. And while other drinks companies were also becoming enormously profitable, Coca-Cola established a presence like no other. Its name and logo began to appear in all the major locations where Americans enjoyed themselves, and it was sponsored by the sports and movie stars idolised by children and young people. The association was established between the drink and enjoyment, between a bottle or can of Coke and a happy childhood experience.

The world at large was also fast becoming Coca-Cola's market. By 1971, more than one half of the company's profits were generated outside the USA, although some locations proved resistant. In some regions – the Middle East, Africa and South-East Asia – the company had to invest in new water facilities for the bottling plants. Branching into costly desalination and hydrological ventures cost the company dear. Despite this, the profits reached dizzying heights: profits of $137 million in 1954 rose to $2.58 billion thirty years later.[15]

Throughout these years of massive expansion, Coca-Cola benefitted from US aid programmes which funded overseas projects, receiving assistance to grow sugar and citrus fruits and to establish new bottling plants. Millions of dollars came the way of the Coca-Cola Company to develop projects in the Caribbean, Africa and Asia. In order to improve poor local water supplies, Coca-Cola persuaded the US Government to support the company's drive to create bottling facilities worldwide. Federal money and guarantees, allied to a new generation of ambitious corporate executives zealous about selling soft drinks around the world, proved a potent mix. Despite failures and imperfections, the resulting water schemes brought clean water to many poor regions for the first time, but the schemes also proved a huge advantage to the soft-drinks industry.

In places troubled by water shortages, though, the Coca-Cola company found itself embroiled in political and legal conflict. Bottling Coca-Cola required massive volumes of water and local objectors sometimes even managed to disrupt and halt production. In a world increasingly alert to the need to garner and preserve global water reserves, the soft-drinks industry had become embroiled in one of the world's major environmental struggles. In a curious twist to this story, it was water that came to their commercial rescue.[16]

Beginning in the 1980s, soft-drinks companies benefitted from the astonishing new demand for bottled water in the USA. Odd as it sounds, the growth of the bottled water economy also boosted the sale of soft drinks. In part, this was related to the decline of old civic amenities. Traditional US civic and urban amenities, especially water supplies, were crumbling, and a number of major scandals surrounding polluted and poisoned water supplies played into the hands of the drinks companies. Their products – new bottled water or fizzy drinks – seemed safer and better than tap water. As the average US per capita consumption of tap water dropped, people began to turn to the products of the major drinks companies.[17]

After a great deal of hesitation (and prompted by the example of Pepsi-Cola), Coca-Cola finally launched its own brand of bottled water in 1999. It was an immediate, dazzling success which almost defies belief. Soft-drinks companies bought water from municipalities at a tiny cost per gallon, bottled it while adding a degree of mineral salts, and sold it at $4.35. Not surprisingly, those same companies also campaigned against tap water, disparaging it in the mass media.[18]

This was the origin of a massive global industry. An endless variety of bottled waters have transformed the way many of us drink. By 2013, the global market for bottled water was $157 billion, and is expected to reach $280 billion by 2020. In the UK alone, the retail value in 2015 was $2.5 billion. And all for a substance that simply falls from the sky.[19]

* * *

Though millions of people around the world have turned to American soft drinks since the Second World War, the most dramatic success of those drinks is to be found where they first began – in the USA. There, for more than a century, the soft-drinks business had been built around a plentiful supply of cheap sugar. Sugar had been the central ingredient in the popularity of fizzy drinks but, by the late twentieth century, it was clear that people's consumption of ever more fizzy drinks, in league with changing food and eating habits, was having a disastrous effect on their physical well-being. In the USA, sweetened soft drinks had been transformed from an infrequent treat to a persistent daily habit. In the 1950s, the annual per capita consumption of highly calorific soft drinks had been 11 gallons; fifty years later, that had risen to an astonishing 36 gallons. In the process, Americans were consuming 35lb of sweeteners each year from soft drinks alone.

Sugar had become a major problem. Throughout much of the twentieth century, sugar had been a matter of political and economic dispute in the USA and, in 1974, the old sugar quota system, designed to protect US sugar interests and control sugar prices, was ended by Congress. The drinks and confectionary lobby hoped this would give them access to still cheaper sugar, but the opposite happened. At first, sugar prices increased dramatically, then fluctuated up and down wildly. By the late 1970s, the US sugar industry was keen on a return to the stability provided by federal protection. For their part, the soft-drinks companies were tired of their reliance on volatile world sugar prices and supplies, and began to look for alternative sweeteners. The answer was close to hand.[20]

The search for artificial sweeteners had begun in the late nineteenth century, driven by pharmacists experimenting in their labs. Saccharin, for instance, had been discovered in 1879, and was sold commercially after 1914, having thrived in the years of wartime sugar shortages. Another major sweetener – cyclamate – emerged after 1945, and various combinations of these products appealed to the soft-drinks companies, especially from the 1950s onwards, in their early efforts to produce a reduced-calorie version of their main products. However, health concerns about artificial sweeteners in the 1960s and '70s – and an increased scrutiny by the US Food and Drug Administration (FDA) into such products – lent a new urgency to the search for safe, new sweeteners. The solution seemed to be NutraSweet (aspartame). The makers of NutraSweet realised that the product's future was linked to the soft-drinks market. Profits boomed and the makers became part of the massive Monsanto conglomerate.

By the end of the twentieth century, with a rising crescendo of concern about artificial sweeteners and health, and about the impact of sugar on global obesity, the corporate and legal in-fighting about artificial sweeteners became fierce – not surprisingly, perhaps, given the value of the global market for sweeteners. The annual sales for the market-leader – aspartame – stood at $3 billion; the next twelve leading products brought the total value to $3.5 billion.[21] The revolution in American sweeteners came, however, from an unlikely corner of American agriculture.

Like the US sugar industry, US corn was highly protected, regulated and subsidised, with the origins of federal intervention lying in the Depression era of the 1930s. Farmers in the American Midwest produced more and more corn, partly by using new, scientifically developed strains, and partly by innovative, highly mechanised farming systems and equipment. The end result, by the late twentieth century, was that US granaries were full to overflowing. The nation's farmers had produced much more corn than the US could possibly consume.

Extracting a sweetener from corn had long been familiar to agricultural scientists and, even by the late nineteenth century, a number of corn-based sweeteners were available on the US market, and a number of companies specialised in corn syrup. But the taste was never quite right. Then, in 1957, scientists hit upon a process for making high-fructose corn syrup (HFCS). At first, it was more expensive than sugar, but all that changed thanks to US legislation. The 1973 Farm Bill devised a payment for farmers allowing them to grow as much corn as they wanted, while guaranteeing them a profit via federal subsidies. HFCS then became cheap, and a new process made it even sweeter than cane sugar. First the USA, and then the global demand for market sweeteners was utterly transformed.

Coca-Cola, ever cautious about tinkering with its major product, experimented with corn sweeteners in some of its lower-profile drinks. They discovered that customers did not complain and, in 1980, the company switched from cane sugar to HFCS. In 1985, corn syrup became the sweetener for all the company's major drinks in the USA. As usual, the country's wider confectionary industry followed their lead, and corn syrup quickly became the dominant sweetener in the USA. By the mid-1980s, most manufacturers of soft drinks had switched completely to HFCS, and it is only recently that researchers have exposed its possible dangers to health. That sweetener soon infiltrated the wider confectionary and food market, and was being used in a wide range of products, from ketchups to cookies, from cakes to candies.[22] The impact was astonishing.

Best of all, from the companies' viewpoint, the new sweetener dramatically slashed production costs for soft drinks. The Coca-Cola Company also implemented a critical marketing change that was to have equally far-reaching results – they increased the size of their bottles and cans. While the cost of producing their drinks had fallen substantially, the company charged only a few cents more for a much larger volume of drink. Because corn syrup was cheap, 'it paid to go big'. The sizes of servings of soft drinks rose: first to 12fl oz, then 20fl oz containers, and even to 64fl oz 'buckets'. All this went hand in hand with Coke's great ally McDonald's, which had 14,000 outlets in the 1990s, and their own invention of 'supersizing' of their own food products.

In the 1950s, McDonald's served only one size portion of French fries. In 1972, they offered 'large size' and, in 1994, 'supersized fries'. They even promoted their products with the phrase 'Supersize It!' and the pattern was adopted by rival fast-food chains.

In the 1980s, the companies launched their supersized products (20fl oz bottles with fifteen teaspoons of sweetener; one litre bottles with twenty-six teaspoons; and even a 64fl oz version with a massive forty-four teaspoons). As these servings got bigger, children were drinking more and more. By 1995, two out of three American children were drinking a national average of 20fl oz every day. And at its height, a large serving of Coke contained 310 calories.[23]

The end result – in addition to rising profits – was a massive increase in per capita consumption of soft drinks from 28.7 gallons in 1985 to 36.9 gallons in 1998. HFCS now represents 50 per cent of all sweeteners consumed in the USA. It had also become the focus on its own increased medical and scientific scrutiny, with concerns raised about its impact on a range of health issues.

What lies behind these statistics about diet and drink was little less than a human revolution. Americans had begun to consume many, many more calories than they needed. In 1950, the per capita consumption of calorific sweeteners was just over 100lb. Thirty years later, it was over 125lb; by 2000, that had reached 153lb. Bizarre as it seems, the link between this and American agriculture was inescapable – America's farmers were being subsidised to fuel 'an unhealthy trend towards overconsumption of carbohydrate-rich sweeteners'.[24]

What made this trend all the more potent and far-reaching were the major social changes taking place in the nature of US society itself, especially in employment and residential locations. Americans had become a nation of city-dwellers, and much of America's employment was non-manual and sedentary, with 80 per cent of America's city workers employed in service industries. An increasing proportion of the labour force worked in less labour-intensive occupations which required fewer calories than earlier generations. One simple example is the daily commute – fewer people walked to work. In the last forty years of the twentieth century, the number of Americans driving to work increased from 40 million to 110 million, and the average return journey took fifty minutes. In 2003, less than 20 per cent of Americans had some form of daily exercise. As the people of the USA became more sedentary, their food became cheaper and they were consuming ever more calories.[25]

The irony behind all this was that the process was highly subsidised. Soft and fizzy drinks were cheap and very sweet – courtesy of the American taxpayer. Billions of tax dollars ($5.7 billion in 1983 alone) went to subsidise corn production, so American obesity was itself subsidised by American taxes. In 1971–74, only 14 per cent of Americans were obese. By the mid-1990s, that had risen to 22.4 per cent and, by 2008, more than one third of the US population was obese. The formula seemed simple: 'Americans were turning excess sugar into fat.'[26]

American obesity was clearly linked to the consumption of highly sweetened soft drinks. This pattern was, of course, spread unevenly throughout the population. Minority communities – especially the poor – were disproportionately obese. A cheap hamburger and a fizzy drink was often the only affordable way of eating in many communities where incomes were low, welfare high and modestly priced food outlets distant or inaccessible. Nor was this a uniquely American pattern. The dietary revolution taking place in the USA was to be seen in all corners of the globe, and the impact on global health was massive. Millions of people all over the world were consuming increasing volumes of mass-produced, processed drinks and food – all saturated with excessive amounts of sweeteners – and many millions of them were getting fatter. Sugar, once a luxury, then a necessity, had now become the enemy.

* * *

Coca-Cola's remorseless promotion of its product, especially among the poor at home and in less developed nations, finally even alienated one of its senior executives. Jeffrey Dunn, once President for North and South America, spoke openly about the doubts he – and many others – now had about the impact of carbonated drinks (and industrialised foods) on health. The company was directing massive efforts towards selling more and more drinks to those who could ill afford them in the USA and around the world, who had barely enough resources to procure the basic nutritional needs for a normal life. So powerful had their advertising become that they could persuade the less well off, at home and abroad, to buy Coke at the expense of more vital commodities. The poor were staying poor – yet they were also getting fatter. It was a remarkable historical upheaval. For centuries, it had been the rich who tended to be overweight; the indulgent wealthy had been portrayed as obese. Now, the reality had been turned on its head; the poor were becoming the fattest people on earth. Jeffrey Dunn had no doubts about the correlation between rising obesity and the per capita consumption of sugary soft drinks.

The importance of sweetness had been confirmed even in the fierce rivalry between the two soda giants – Coca-Cola and Pepsi-Cola – in the 1980s. They fought each other to a standstill, one sometimes overtaking the other, one claiming to be sweeter or more delicious than the other – but both were carried along by a rising tide of consumption. Despite the rivalries, both companies thrived and sold more and more drinks to their fans. It didn't seem to matter what they said about each other – both sides thrived. And both thrived on selling their sweetened drinks.

When the Coca-Cola Company switched from more expensive refined sugar to the cheaper, high-fructose corn syrup in 1980, their profits rose even higher. So, too, did the marketing budget. By 1984, it reached $181 million. The drive by Coca-Cola was to persuade people to buy ever more Coke, and it worked. By 1997, Americans were drinking 54 gallons of carbonated drinks a year – Coke controlled 45 per cent of the overall market – and sales rose to $18 billion. But because Diet Cokes accounted for only 25 per cent of the sales, people were overwhelmingly drinking sugary drinks – more than 40 gallons a year, or 60,000 calories, equivalent to 3,700 teaspoons of sugar per person.[27]

The consumption of Coca-Cola seemed a perfect confirmation of the 'Pareto Principle' – the theory that 80 per cent of consequences stem from 20 per cent of the causes. In this case, 80 per cent of Coke's consumption was accounted for by 20 per cent of the population. More alarming still, however, that 20 per cent was located at the lower end of the social scale, among the poor and dispossessed who could ill afford to spend rare resources on a drink that added little nutritional value to their diet. Yet the company's marketing policy was to persuade those very people to drink more.

The other market targeted by the companies was the young – the people who would become lifelong Coke drinkers. Although the company developed a policy about not advertising to under-twelves, there were many ways of stimulating an interest in a drink without direct TV appeals. The name, logo and images of Coca-Cola were ubiquitous in the very places where children spent much of their formative leisure time. That and, via careful research and marketing, locating branded drinks dispensers and outlets where the young were likely to shop. The drink was to be placed at the most strategic positions, in corner shops and supermarkets, to persuade people to buy impulsively.[28]

Behind this lay a forensic analysis of shopping habits, market and social research into the shopping and consumer habits of the US population, by every conceivable category – rural and urban, socio-economic class, age, gender and ethnicity, and so on. They were thus able to reach their target groups not merely in the nation's major supermarkets but, critically, in local convenience stores. There, too, drinks were located precisely to catch the eye and the cash of customers – children, say, at nearby schools. The corner shops of the late-twentieth-century USA therefore became lucrative enterprises, their profits flowing primarily from the sweetened drinks and snacks loved by their young customers. The results led to a proliferation of corner stores – and an ever-rising flow of soft drinks and snacks out of their doors in the hands of ever-younger customers.[29]

It was here, in the convenience stores, that drinks companies were developing a powerful loyalty for their brands among the nation's young – get them young, and you had them for life. It was a revival and confirmation of the principle first advocated by Robert Woodruff decades earlier. For critics, as if all this were not bad enough, all the major manufacturers of soft drinks and snacks were making powerful incursions into much poorer countries, nations that were, in some cases, developing rapidly, but nonetheless had swathes of their people suffering serious deprivation. To reach those people, the major international companies began to produce small versions of their drinks and food, offering drinks and snacks that were significantly cheaper because they were sold in smaller sizes.[30]

* * *

Sugar, then, has been at the very heart of the long history of American soft drinks. In the post-war years, it had sweetened the powdered fruit drinks which Americans had mixed with water and served to their families. At their peak, those powdered drinks brought in $800 million. Towards the end of the twentieth century, fruit flavours were added and children were targeted via leafleting and junk mail. When the same drinks were repackaged and sold in small cartons, the drinks became hugely popular, all supported by images of health and nutrition and, above all, they were fun. But food scientists were also at work, devising new fruit flavours, and finding sweeteners. The answer was pure fructose, which was much sweeter than sugar itself. Once the imperfections had been eliminated, food manufacturers could turn to fructose in their products and claim it was good for you. This happened when sugar was under serious attack as the source of worrying health issues, and played into the food industry's hands. Pure fructose seemed to be the answer to the growing band of critics of sugar. It was to take another decade or more before further research began to suggest that sucrose and corn syrup were also likely to encourage health problems, notably heart disease. Today, with the scientific jury still pondering the issue, fructose is seen by many critics to be as dangerous as cane sugar itself.[31]

Other sources of sweetness were, however, also available to the food manufacturers. 'Fruit juice concentrate' emerged by the beginning of the twentieth century as another powerful weapon in the ongoing battle to find sweeteners that were commercially viable and safe. Here, once again, science and advertising concocted another alchemist's dream – until tested in the courts. We now had another sweet additive that, in many cases, had been stripped entirely of its nutritional value.

Sweetness was the essential hook used to target and catch American children. There seemed little that the major corporations would not do to secure a child's loyalty to the sweet products pouring from American factories and into the aisles of supermarkets and corner shops, and ultimately into the hands of the young. The evidence of the success of the food corporations was to be measured, however, not merely in the financial returns of the companies involved, but in the expanding waistline of the American people. As the food corporations grew fat on their sweet products, their customers simply grew fat. And so, too, did millions of consumers the world over. What the USA had pioneered, they exported – sweet-tasting soft drinks and unprecedented levels of obesity.

What could be done about it? One stock reply, a mantra of defenders of the food industry, was that individuals had a choice – no one was *forced* to buy sweetened food and drink. Consumers could resist the blandishments of the modern diet, and look to their own physical well-being. Millions took this route of self-control and, over time, there emerged another rival industry – that of personal fitness, diets, gyms, food fads and bogus claims – much of it in pursuit not merely of a healthy body, but of an idealised body. For all that, it was obvious that more, much more, was needed to turn the tide of global obesity.

16

Turning the Tide – Beyond the Sugar Tax

THERE HAVE BEEN a bewildering assortment of reactions to the growing awareness of the problems of obesity. International organisations – notably the World Heath Organization – have launched global initiatives against excessive sugar, while a number of individual governments, feeling the strain on their medical facilities, have turned to the idea of a 'sugar tax' on sugared drinks, which seem to be the main culprit.[1] At an individual level, millions of people have set out to counter, in their own distinctive fashion, the problems of obesity and inactivity by following regimes of diet, exercise and the pursuit of general physical well-being. In the process, new industries have emerged to cater for their needs – healthier foodstuffs and shops, diets and exercise programmes, costly fitness centres – a whole culture has been developed as an antidote to the forces that have produced worldwide obesity.

Diets now abound of every conceivable kind and variety, some sensible and medically approved, others rooted in ever more esoteric – and downright dangerous – philosophies and beliefs. Specialist magazines, diet books by the yard, TV programmes, specialist outlets and shops, regimes, diet pills, gymnasiums – all and more offer themselves as an alternative lifestyle and counter-balance to the world of sweet obesity.

The arguments about obesity have even spilled over from medicine and nutrition into a wider social discussion about environmental issues and even about the future of the planet. It also formed an important turning point in the history of the women's movement when Susie Orbach's book – *Fat is a Feminist Issue* – focused attention on weight and body image as key issues in the question of gender politics.

Standing in sharp contrast to obesity, then, is the opposing cult of athleticism and the pursuit of a healthy, ever more beautiful body. Streams of people of all ages, shapes and sizes run, jog, walk, swim and cycle – all bearing witness to a determination to counter obesity and its threat of ill health. Gymnasiums dot the urban landscapes, many of them with plate-glass windows which allow lesser mortals to glimpse the activity within, and wonder at the sweat being expended on the vast array of costly machines designed to keep body and soul in a more acceptable shape. And, of course, to keep weight at an acceptable and healthy level.

In 2015, an estimated one person in eight in Britain – 1.5 million of them in London alone – belonged to a gym. There were an estimated 6,312 'fitness facilities' around Britain, and their market value stood at an estimated £4.3 billion. The phenomenon is now openly discussed as 'an industry', its supporters and investors relishing the healthy state of their financial returns year by year. In the USA, naturally, the figures are even more striking. The current estimate of about 36,000 facilities generated an annual income in 2016 of $25.8 billion, and were used by 55 million members. The sporting equipment sold to such places amounted to $5.12 billion. Worldwide, an estimated 151 million people used such fitness centers in 2015.[2]

The evidence for the worldwide diet industry is, if anything, even more astonishing. The British diet industry alone is now worth £2 billion a year, at a time when the NHS spends £2.33 billion on accident and emergency services. In the USA, the industry is worth an annual $60 billion. But we need to consider this alongside some financial snapshots from the fast-food industry. The advertising budget for McDonald's, for instance, is $2 billion. Another $3 billion is spent each year on packaging breakfast cereals aimed at children.

These figures provide a glimpse into the colossal financial and commercial forces at work around us. On the one side, food and drink industries, and their vital advertising allies; on the other, a huge medical alliance and its supporters – the people who have to deal with the problems of obesity – in league with a fitness lobby offering commercial alternatives to obesity. So great is the problem, so powerful the commercial forces at work behind the question of obesity, that even governments hesitate over how best to tackle the problem.

It was clear enough by the early years of the twenty-first century that obesity was yet another consequence of the wider impact of globalisation. In this case, it was a universal damage wrought by modern food and drink, all disgorged by the world's corporate giants. Yet the problem manifested itself more directly – and worryingly – at the national level; it was up to individual governments to find solutions to their own national problems of obesity. Their reactions varied from country to country. In the case of England, the matter came to a head with the publication in 2015 of a major report by Public Health England: *Sugar Reduction: the Evidence for Action*. The report pulled no punches, offering a withering review of the problem and a stark outline of the evidence. Along with many other countries, England found itself plagued by obesity on an unprecedented scale, with all the familiar health threats.

The 2015 report blamed sugar. The opening sentence set the tone: 'We are eating too much sugar, and it is bad for our health.' The evidence to support the case was stark; almost 25 per cent of adults, 10 per cent of 4–5-year-olds and 19 per cent of 10–11-year-olds in England were recognized as obese, with significant numbers also being overweight.[3]

The report was, however, merely the latest episode in a long process of social and political concern about obesity. There had been a rising crescendo of public alarm throughout the early years of the twenty-first century – government reports, medical enquiries and a veritable blizzard of articles and programmes in the media, most notably, perhaps, by celebrity chefs such as Jamie Oliver. All had served to push the question of obesity to the centre of public and political attention. It was no longer something that could be overlooked.

At one level, the problem was indisputable, because it was so visible. The evidence of obesity in society is plain for everyone to see, every day, in public life. It was, however, medical professionals who found themselves struggling to cope with the consequences of obesity. Yet the root causes of the problem were much less obvious – and debatable – although there was no doubt about the impact on the NHS. The 2015 report asserted: 'Obesity and its consequences alone cost the NHS £5.1 billion per year . . .' and claimed it had no doubts about the prime cause of the problem – the high concentrations of sugar in the nation's food and drink.[4]

The short-term explanation lay in the profound changes that had transformed our relationship with food and drink in the years after the Second World War. For a start, in real terms, food became cheaper than ever. But food itself was also *different* – much of it became processed and industrialised, with sugar used extensively in that process. It was also marketed and sold in a very different manner, overwhelmingly in supermarkets. At first glance this might seem a marginal issue in any discussion about obesity, but these new forms of shopping played a central role in the complex transformation of people's food and drink. Supermarkets were a vital agent in infiltrating unprecedented volumes of sugar into people's diet.

These years also heralded a new age of mass consumption, and people were constantly urged, by advertising, promotions and slick propaganda, to buy and consume more of *everything* – including food.[5] In common with material life at large, we simply consumed more and more. Just as we had been lured into filling our lives with an abundance of material objects we once managed to live without, so had we been wooed into eating more food than we needed. And much of that food was of little nutritional value and had been sweetened en route to the supermarket shelves. The end result is that sugar is being consumed, largely unknowingly, on a scale that would have made an eighteenth-century sugar planter rub his hands with delight.

The 2015 study by Public Health England paid particular attention to children. The evidence confirmed that children were consuming on average three times the medically recommended levels of sugar, and adults twice the levels.[6] The main sources of that sugar were as might be expected – soft drinks, household sugar, confectionery, fruit juice, biscuits and similar treats, and breakfast cereals. For adults, alcohol was a major source of sugar.

There were some variations between different age groups. Among teenagers, for example, soft drinks provided the largest source of sugar, while younger children absorb sugar via biscuits, cakes, breakfast cereals, confectionery and fruit juices. Again, and not surprisingly, perhaps, sugar intake was greatest among low-income groups, and consequently the related problems of obesity – among all ages – were worst in communities and areas blighted by widespread deprivation.

Stated crudely, the poor, the unemployed – indeed, whole communities abandoned by the collapse of local industries in the face of globalisation – once again fared worst. Yet it is hardly original to restate that the poor – on both sides of the Atlantic – suffer most from the personal and physical problems created by a bad diet of cheap, sweet, processed foods and drinks.

Here, then, was an English perspective on the much broader problem which linked a nation's food to a wider study of how people were persuaded to buy the food and drink they consume. Lurking over the entire question is the power of modern advertising, and the way food and drink are marketed by costly and manipulative campaigns. The changing dietary habits of, say, the past forty years are closely related to the way food and drink products are promoted. Advertising is no longer the world of colourful public hoardings, or simple catchy TV adverts and jingles. The very channels of advertising have themselves undergone a revolution to keep pace with the emergence of the Internet and social media, notably in the form of 'pop-up' ads on smartphones, tablets and computers. Recent restrictions on advertising aimed at children on TV and in supermarkets are easily circumvented by relocating adverts to social media. The more children use smartphones or tablets, the more adverts they are likely to see for sweet food and drink. And the industry for online advertising is growing all the time: in 2013, £6.3 billion was spent on advertising on the Internet in the UK.

A great deal of the advertising of food and drink is aimed specifically at children. They are bombarded with cartoon characters from favourite shows, colourful images and stories, and child-friendly packaging, and many adverts are designed both to entertain the young and win them over to a particular product. Away from the screen, elaborate displays of sweets, chocolates, cakes and drinks are strategically located in supermarkets to catch a child's attention in the hope of an impulsive purchase by a parent or by children themselves. It is these products which usually contain unhealthy levels of sugar. Recent research confirms that such advertising, both in the new media and on TV, is successful in persuading children to crave, then to buy, sugary items – children will consistently opt for sweet products.[7]

We also know that children's choices can be swayed by a product promoted by a popular star, typically a famous sportsman or woman, and this became a well-used commercial tactic. So, too, is the ploy of proclaiming price cuts, or offering 'two-for-one' special offers. Sales are dramatically affected by such offers. We also know that goods with a high sugar content are the very items more likely to be placed on offer, because, again, market research has confirmed that people are tempted to buy goods of high-sugar content in this way. Equally, the exact location of a product in a supermarket is important; which spot in the aisle or on the shelf is most likely to attract a child's attention. The end result is that an estimated 7–8 per cent of *all* the sugar entering the home is bought in this fashion – by eye-catching price reductions or special offers.[8] These marketing techniques have been tried and tested, in one form or another, for half a century and no serious student or researcher in the field is in any doubt that they are highly successful in selling sweet products.

While sugar is the main dietary culprit behind modern obesity, it has been able to reach its current levels of consumption via the skills and power of marketing agencies.[9] The achievements of advertisers are, in their turn, closely linked to the recent history of changing shopping habits, none more dramatic than the rise of the modern supermarket and the emergence of other food outlets on the high street, in the workplace and school. There is, then, a complex, interconnected web of factors which determine how and what people buy to eat and drink – but the focus returns, time and again, to sugar.

* * *

Our diet has changed with astonishing speed, but it now embraces a range and choice of foods available which is more profuse, varied and more international than our grandparents could possibly have imagined. We now find ourselves surrounded by countless opportunities to buy food: supermarkets, revitalised corner shops, restaurants, fast-food outlets, takeaways, cafés and coffee shops. Food and drink outlets in the UK now account for a large and growing proportion of all the meals consumed – 18 per cent of meals came from this sector in 2015, and 75 per cent of the population claimed that they ate out or bought takeaway food in

2014.[10] The food sold at such outlets are the very foods in which sugar levels are at their highest and, here again, sugar makes its presence felt, often in unlikely places. The massive expansion of 'coffee culture', for example, has also encouraged extra sugar consumption via its proliferation of new forms of hot beverages, alongside all the varieties of sugary confectionery that are offered with the hot drinks. Customers came to expect not simply a coffee but a variety of sweet delights at their local coffee shop. Even the coffee changed – heavily sweetened, flavoured syrups can now be added to a hot coffee, and often contain as much sugar as the more obvious cans of fizzy drinks.[11]

This proliferation of eating and drinking places, and the availability of pre-cooked or takeaway meals, form part of a broad cultural invitation to eat more and more food. We also eat more because food is cheaper than ever. Today in the UK, around 15 per cent of most people's weekly expenditure goes on food; half a century ago, it was 33 per cent. In addition, the portion sizes we consume – especially in takeaway and prepared foods – have grown. The end result is that the average Briton consumes between 200 and 300 more calories each day than the body requires.[12] And the finger once again points at sugar. But what is to be done?

All the evidence, by 2015, confirmed that any attack on sugar and obesity needed to move well beyond mere exhortation or providing educational and health messages about diet. That had been tried many times, but the nation – the world – was still growing fatter. The target audience seemed deaf to such pleas. Poor people especially – the very people with limited access to healthy food – persistently showed themselves unmoved by sermons about healthy, sugar-free diets.

The key targets in the attack on sugary diets are easily identified: the manufacturers who introduce high levels of sugar into their products, the advertising industry which supports them, and the powerful supermarket chains which tantalise customers – especially the young – with their irresistible displays and targeted price cuts. By 2015, there was broad agreement about the urgent need to reduce the volumes of sugar added to food and drink. It was equally important to curtail the aggressive marketing of sweet items at children. There was an encouraging example to follow: the British campaign to persuade food manufacturers to reduce salt levels in food. Salt in bread, for example, had been reduced by more than 40 per cent since the 1980s, without any appreciable customer backlash.[13] This raised, though, an unresolved and fundamental question about the different biological and physiological reactions to sugar and salt. Humans seem to have an innate love for sweetness, but not for salt. In simple terms, we don't crave salt in the same way as sugar, and we don't miss salt in the way we love and would miss sugar.

By the time *Sugar Reduction* was published in 2015, what had emerged from this welter of evidence and argument – not only in Britain, but globally – was a need to make decisive moves against sugar. One apparently simple and very tempting proposal was the idea of a 'sugar tax'.

Naturally enough, the prospects of a tax on sugary items raised howls of outrage from the sugar lobby, from drinks and food manufacturers and from the retail outlets which sell those products. However strong the evidence about the links between obesity and sugar, the food industry was unwilling to see its trade limited by taxation. Two main factors, however, lent support to the idea of a sugar tax when it was first raised in

Britain. First, it had already been pioneered in a number of other countries, all of them alarmed about their own problems of obesity and sugar consumption. Secondly, the early evidence suggested that such taxes seemed to work. Various forms of sugar taxes had been introduced in Norway, Finland, Hungary, France and Mexico, and in some US cities. And they seemed to be having the desired effect. Sales of sweetened soft drinks, for instance, had fallen more or less in line with the percentage of sugar taxation; the 10 per cent Mexican tax has seen a fall of 6–9 per cent in soft-drink sales, a fall most striking among the poor, who were the most at risk from spiralling ill health.[14]

The British Conservative Government was, by instinct, resistant to interference in industry and uneasy about new forms of taxation, but the 2015 report forced its hand. Following a number of delays, and various leaks, the report, which had been commissioned by the Government in the first place, finally appeared in 2015. The British media then took up the case with a vengeance.

Newspapers of all political persuasions launched an attack on sugar. *The Times* – traditionally a Conservative advocate – thundered its own vocal support for a sugar tax, and chided ministers for holding back both the evidence and the necessary political action. In a leading article in October 2015, *The Times* neatly summarised the argument for such a tax. It offered its own résumé of the evidence, lending publicity to the problem of highly sweetened food and drink, the rise of obesity – more especially among children – and the financial stress on the NHS.

The Times also pointed to a precedent. Between 1994 and 2003, a number of medical and governmental studies had urged a reduction in the volumes of salt added to processed foods. A public-awareness campaign against salt – which is still ongoing – had gone hand in hand with discussions with food manufacturers to scale back salt in foodstuffs. *The Times* argued that 'there is no reason why sugar cannot undergo a similar trajectory'. While accepting that individuals needed to make their own informed decisions about what they buy and drink, *The Times* realised that the time had long gone when personal choice might slow the rise of obesity. The time was ripe for the Government to 'give serious consideration to a sugar tax'.[15]

Following the report's publication in October 2015, a number of journalistic heavy-hitters joined in the attack on sugar. They, of course, were unrestrained by the protocols of the leader writers on *The Times*. Some took a much more severe line, arguing that a sugar tax alone was inadequate; it needed to be bolstered by a ban on fast-food outlets in railway stations, airports and other public places. In answer to critics who found such ideas unnecessarily draconian and intrusive, they made the obvious – and telling – point that critics had raised very similar objections over the past thirty years when the attack on smoking in public places had gathered momentum.[16] Jamie Oliver, the celebrity chef with an enormous TV following and a thriving commercial empire of restaurants and books, also caught the public eye by linking tobacco and sugar, calling sugar 'the next tobacco'.[17] Given his celebrity status, Oliver's ideas and pronouncements received major coverage.

In fact, medical authorities had already been making much the same point – that we should think of the threat from sugar much as we once thought of tobacco.[18] Epidemiologists, medical experts of various specialisms and medical sociologists, all and more had for some time been pointing to the similarities. What was once said about tobacco was now being alleged of sugar. This ever-increasing group of critics, on both sides of the Atlantic, were growing in numbers and stridency, and formed a broad if unorganised coalition of famous media faces, medical experts, social scientists and politicians. Each, from their own corner, accused the food-and-drink industry of ignoring the widespread damage caused by sugar. Sugar's opponents saw the food industry cynically enhancing profits with little or no regard for the physical and social impact sugar was having – especially among the young.[19]

Linking sugar and tobacco, and comparing the behaviour of the sugar lobby to the rearguard tactics of the tobacco lobby, proved to be the most telling of blows. By the early twenty-first century, no one could be in any doubt of the ravages caused by tobacco, and to place tobacco alongside sugar was perhaps *the* most damaging blow to sugar's credibility. It was now up to the sugar lobby to prove that the tobacco–sugar link was inaccurate or plain wrong.

Some voluntary agreements had already been brokered by the British Government with companies to reduce the sugar contents of their food and drinks. But such deals inevitably involved only willing partners, and were still a long way from addressing the core problem. Demands for a sugar tax would not go away.

The first practical step towards such a tax was taken by the NHS itself, with a proposed tax on sugar and sweet food and drinks throughout all NHS infrastructure – in hospital catering locations, shops and staff facilities.[20] By the New Year of 2016, the British Cabinet, despite its earlier hard stance, had been won over to the idea of a sugar tax as a viable initiative in the fight against obesity. It was agreed to introduce a tax on sugary drinks in 2018. It had been a hard sell, but ministers were apparently persuaded by the evidence from countries which had already levied taxes on sugary confectionery and drinks. As documented earlier, in Mexico, the tax on soft drinks had seen a marked fall in sugary drink consumption, and this despite Mexicans having become infamous as a people who loved soft, sugary drinks.[21] Norway's tax had encouraged people to eat sweets and chocolates less frequently and, in Finland, soft-drink sales had fallen after a sugar tax. In Hungary, there had been a sharp decline in heavily sweetened products, with companies manufacturing items containing less sugar to avoid the tax.[22]

Through all this, the soft-drink industry, in particular, had not remained idle. Faced with this growing alarm about sugar, and by the upsurge of concern among politicians and consumers, they responded. The buzzwords were 'reformulation', smaller packaging sizes, and a growing emphasis on low- or no-calorie options.

So it was, in the summer of 2016, that millions of TV viewers found themselves bombarded by adverts proclaiming Coca-Cola to be 'Zero Sugar'. Sugar was about to be taxed and had, now, been virtually 'banned' as an ingredient from one of the products that it had done so much to help become pre-eminent the world over.[23] Sugar had taken on a pariah status. This was little short of a major transformation in the way sugar was used and perceived, and how it was now seen by the food industry and by consumers. What had, for centuries, been promoted as an ingredient that delivers simple pleasure – a commodity that made food and drink tastier and made us feel happy – was now being denounced for its capacity to do untold harm.

Conclusion

Bitter-Sweet Prospects

How did it come to this? How did so many millions of people become so overweight? And how has obesity managed not only to grab headlines the world over, but to become a pressing concern for governments and health agencies everywhere?

Through the ages, human history has *always* experienced instances of overweight people, largely caused by a number of well-known medical conditions. And, as we have seen, being overweight in the past traditionally meant being a target of ridicule and personal abuse. This still remains the case, with those who suffer with excessive weight continuing to complain of their treatment by society in general, but the issue is now being played out at a whole new, unprecedented level.

The current problem of obesity has produced an unusual convergence of opinion – the coming together of a broad coalition of individuals and groups anxious to tackle what they see as a major health problem. An array of medical experts, social commentators, media analysts, politicians – and, not to be discounted, parents anxious to shield their offspring from the patterns of behaviour that seem to lead inexorably towards being overweight – all these have come together, first to complain about the problem and then to do something about it.

But what exactly are they complaining about? If people eat and drink unhealthy food and drink, that, surely, is *their* choice, *their* decision? The stark, liberal view would be that people should be free to choose to do, and to conduct their lives, in whatever way they wish. They have a choice – it's *their* responsibility.

The trouble is, the consequences are not their responsibility alone – they are foisted upon everyone who has to pay the enormous cost of treating or caring for the impact of obesity. And the decision about diet isn't merely a simple matter of individual choice. People are steered towards decisions about consumption by powerful – in some respects irresistible – commercial forces. They cleverly blend the appropriate strands of science (food, nutritional and medical research) with the findings of market research and advertising. The entire package is then targeted to capture people at their most vulnerable and suggestive. Scientists have long known that children, from birth, like sweetness; the food industries have created products which satisfy and nurture that taste; marketing executives then devise means of exposing their young target audience to the irresistible temptations of sweet treats. The result has been what contemporary parlance might call a 'perfect storm' – a confluence of irresistible forces which people are unable to withstand. At the centre of that storm lies the role of sugar.

Analysts of the chemistry and physiology of obesity have returned time and again to sugar. Although sugar is not alone in the cocktail of ingredients that have transformed the world's dietary habits in recent decades, it has proved to be the pre-eminent ingredient in a huge range of those foodstuffs that create such damaging consequences for human well-being.

Yet this is curious, because sugar did *not* emerge from the research labs of food scientists, but from a very long historical presence in mankind's diet stretching back millennia. People have traditionally liked sugar, adding it to their drink and food in societies and cultures around the world. We now know that those sweet pleasures eventually came at a price – initially in the form of serious dental trouble. The bad teeth of Elizabeth I, Louis XIV, and working-class children in the late nineteenth century were clear warnings of what was to follow. The scale of the health problems – from royal dental decay in the late sixteenth century to today's obesity epidemic – was, of course, utterly different. But the cause was the same.

To add to the conundrum, mankind's relationship with sugar is a force of considerable power and significance. Moreover, there has been no interruption in that relationship. To put it crudely, the world of modern industrialised foods with their own addictive use of sugar has something in common with doctors in Baghdad in AD 1000: both recognized that people like sweet tastes. And around that desire for sweetness it was possible to wrap a host of things – be it ancient medicine or a bottle of Coca-Cola.

For all that, sugar also corrupts – most obviously and directly through the damage it inflicts on consumers' teeth. The most compelling evidence became available with the emergence of modern dentistry, and the ever more detailed scrutiny of children's teeth in the Western world. It was obvious that those who relied most heavily on a sugary diet – and that meant, primarily, the poor – were the most seriously damaged by sugar. Their teeth bore testimony to the destructive effects of a sugar-rich diet. Although now confirmed by medical science, it had already been visible, among the upper echelons of society, centuries before. In the years when only the rich could afford sugar, and when decorative sugar work were symbols of power and prestige, the rich bore traces of their sweet indulgence in their rotten teeth. European monarchs were the more prominent victims of a passion for sugar. At the time, having a sweet tooth often meant having a rotten one – or sometimes none at all.

Sugar's ability to corrupt, though, went far deeper and wider than dental health. If we stand back from the broad sweep of the history of sugar, and consider how sugar rose from rare luxury to an item of mass consumption, its power to corrupt becomes quite startling. It transformed the physical and environmental face of large expanses of the earth's surface. It was also primarily responsible for one of the most hideous and damaging migrations of humankind in history, which has echoes and repercussions that still trouble us today. For the best part of four centuries, sugar was cultivated in Brazil and the Caribbean by enslaved Africans and their descendants born into slavery. Sugar and slavery went hand in hand, and it seemed to most of the people involved in the system – except, of course, the slaves themselves – that there could be no sugar without slavery. The crudest measurement of sugar's corrupting influence was that the Western world devised, perfected and justified that most brutal of systems for its own pleasure and profit. What greater corruption could there be?

Many millions of Africans were uprooted and shipped in the most debilitating and degrading of conditions thousands of miles – and all for what? To feed the pleasures and palates of the Western world and to profit their masters. The sugar they produced became the sweet delight of millions who knew little (and cared even less) of the slaves' wretchedness. Much the same reaction followed the introduction of indentured labour to the former slave islands, and to new sugar economies the world over. Their efforts enabled sugar to become a viable commercial crop in new tropical settings and become a commodity produced worldwide – from Mauritius to Hawaii. There, as in the Americas, they toiled on plantations, and the plantation model became the chosen way of developing a string of new tropical commodities. But the plantation also wrought enormous environmental damage, as gangs of labourers burned their way through the native habitat to clear the land for the cultivation of sugar. The end result was human and ecological damage on a scale that is hard to assess, not least because it created an utterly new world. What emerged were peoples and habitats which, today, seem natural and timeless. In fact, both had been reshaped by foreign oppressors. And at the heart of those human and environmental changes was the story of sugar.

The ever-increasing volumes of cane sugar, later joined by beet sugar in the nineteenth century, won over the world to sweetness in food and drink. The ancient luxury of the rich was now an everyday essential of the common man, bringing pleasure and energy to labouring people around the world. The by-product of sugar production – rum – did much the same, though it, too, inflicted its own corrupting influence on peoples such as those indigenous to the Americas.

By 1900, sugar was cultivated and produced all over the world, and had become an essential item in the diet of millions. So valuable were the sugar-producing regions that the USA wielded its influence to ensure American control over sugar supplies. Like Britain and France in the eighteenth century, the USA in the early twentieth century saw sugar as an important element in the way it defined its power and strategy. In its dealings with Cuba, the American addiction to sweetness was to have profound consequences on global politics throughout the second half of the twentieth century.

American power came in various guises in the twentieth century. By 1945, it was, most obviously, the world's major military superpower, but its influence spread well beyond its military might. The power of major US companies provided a model of corporate power for others, and there emerged a galaxy of global conglomerates that came to wield unprecedented power the world over. By the early years of the twenty-first century, those corporations – most owing no loyalty to individual states – controlled or dominated the world's supplies of food and drink. And central to many of those commodities lay sugar itself.

As global diets became ever more processed and industrialised, sugar and sweeteners secured their unique importance in the whole process. Cane sugar had fuelled the initial global demand for sweetness, but concerns about sugar, and discoveries of new sweeteners, saw the emergence of other methods of sweetening food and drink. In the process, the world's processed foods and drinks were sweetened to extraordinary degrees, apparently without limit. And the result? Cardiac problems were just one of a multitude of ailments that stemmed from the plague of obesity that began to characterise a vast proportion of the world's population in the early twenty-first century.

No serious medical observer of this drift towards global obesity doubted where the prime cause lay. They blamed the unprecedented volumes of sweeteners liberally introduced to industrial food and drink, and all cleverly promoted by the persuasive cunning of modern marketing. Naturally enough, the sugar lobby, alongside its powerful allies in the food and drink industries and in advertising, put up a stout and often disingenuous defence. The facts were often in short supply; what mattered were the figures, the profits and progress of commercial products, reported to shareholders and board members. So enormous, so global were those corporations that they straddled national boundaries as if they did not exist. Individual nation states – not even the USA – could bring them to heel, although often that was because the corporations had the power to wield undue influence over politicians in the world's capitals, most notably in Washington.

Yet the tide *has* begun to turn, and the clearest sign of that change is to be found in today's newest range of Coca-Cola and in the company's current adverts. 'Zero Sugar' is the new slogan, the current motif emblazoned on company products. That all-powerful corporation is currently promoting one of its major products by highlighting something that it lacks. The drink is, it is now claimed, better and healthier than anything that has gone before it, because it contains *no* sugar. Who would have thought it possible? After all, that drink had been devised and promoted for more than a century on the basis of its distinctively sweet taste. It had hit the 'bliss point' more consistently and with greater commercial success than any other product in human history, and had been swept to global fame and fortune by a unique formula which was sweetened by enormous additions of sugar. Now, in 2016, the game was up for sugar, and Coca-Cola was forced to put a significant amount of its corporate might behind the idea of 'zero' sugar.

That company's dramatic change of approach to the formulation and promotion of one of its high-profile products was important in itself, but it was perhaps even more revealing as a signpost towards the direction in which the food and drink industry as a whole are heading. It is impossible to know how far or how fast that journey will take. But that the world's major soft-drinks company has abandoned sugar for one of its highest-profile products, and proclaimed that fact as a key promotional strategy, forms a decisive, seismic shift. Coca-Cola has traditionally been a commercial pioneer – it has always led the way, and what they do, others quickly follow. Others may also now be ready to consider abandoning sugar for similar products, however much sugar has proven itself to be a cash cow over the years.

In the last few years, it would have been difficult for people most closely involved in the food and drink industry *not* to have noticed the heightened tactics of the opponents of sugar, most critically the shift to the accusation that 'sugar is the new tobacco'. No corporate Board, looking back over the commercial troubles of tobacco over the past half century, can comfortably allow their own sugar-laden products to be placed in the same category as tobacco. The damage, the lawsuits, the commercial destruction of the tobacco industry – all provide an object lesson of what to avoid. Yet sugar is now inextricably linked with tobacco when considering harmful consequences.

No one expects sugar to vanish. It is an industry that employs too many people, and the cultural attachment to sugar runs too deep for sugar to simply disappear. Currently, there are 120 countries producing 180 million tons of sugar. In any case, how do we get round the simple, undeniable fact – people love sweetness and, as we have seen, for centuries have gone to great trouble to enhance their food and drink with sugar. For all the importance of beet and corn sugars, cane sugar continues to account for two thirds of the world's sugar supplies. The world's sweet tooth continues to rely on cane sugar, much as it did four centuries ago. As people have known for millennia, 'Sweetness is the most basic form of tastiness and of pleasure itself.'[1]

Bibliography and Further Reading

Elizabeth Abbott, *Sugar – A Bittersweet History*, London, 2009

Fernand Braudel, *Capitalism and Material Life, 1400–1800*, London, 1967 edn

Jacob Adler, *Claus Spreckels: the Sugar King in Hawaii*, Honolulu, 1966

Cesar J. Ayala, *American Sugar Kingdom: the Plantation Economy of the Spanish Caribbean, 1898–1934*, Chapel Hill, 1999

Peter Brears, *Cooking and Dining in Medieval England*, London, 2008

Linda Civitello, *Cuisine and Culture – A History of Food and People*, Hoboken, 2007

Carole Counihan and Penny Van Esterik, eds, *Food and Culture – A Reader*, London, 1997

David Crawford and Robert W. Jeffery, eds, *Obesity Prevention and Public Health*, Oxford, 2005

Francis Delpeuch, et al., *Globesity: a Planet Out of Control*, London, 2009

Alfred Eichner, *The Emergence of Oligopoly – Sugar Refining as a Case Study*, Baltimore, 1969

Markman Ellis, Richard Coulton, Matthew Mauger, *Empire of Tea – The Asian Leaf that Conquered the World*, London, 2015

Bartow J. Elmore, *Citizen Coke – The Making of Coca-Cola Capitalism*, New York, 2015

David Eltis and David Richardson, *Atlas of the Atlantic Slave Trade*, New Haven, 2010

Ben Fine, Michael Heasman and Judith Wright, *Consumption in the Age of Affluence: The World of Food*, London, 1996

Richard Follett, *The Sugar Masters – Planters and Slaves in Louisiana's Cane World, 1820–1860,* Baton Rouge, 2005

David Gentilcore, *Food and Health in Early Modern Europe – Diet, Medicine and Society, 1450–1800*, London, 2016

Darra Goldstein, ed., *The Oxford Companion to Sugar and Sweets*, New York, 2016

Cindy Hahamovitch, *No Man's Land: Jamaican Guestworkers in America and the Global History of Deportable Labor*, Princeton, 2011

B. W. Higman, *A Concise History of the Caribbean*, Cambridge, 2011

Gail M. Hollander, *Raising Cane in the 'Glades: the Global Sugar Trade and the Transformation of Florida*, Chicago, 2008

Richard J. Hooker, *Food and Drink in America – A History*, Indianapolis, 1981, p. 130

Reginald Horsman, *Feast or Famine – Food and Drink in American Westward Expansion*, Columbia, Missouri, 2008

Colin Jones, *The Smile Revolution in Eighteenth-Century Paris*, Oxford, 2014

Kenneth F. Kiple and Kriemhild Conee Ornelas, eds, *Cambridge World History of Food*, 2 vols, Cambridge, 2000

Rachel Laudan, *Cuisine and Empire: Cooking in World History*, Berkeley, 2013

Hilary Lawrence, *Not on the Label – What Really Goes into Food on Your Plate*, London, 2013 edn

David Lewis and Margaret Leitch, *Fat Planet – The Obesity Trap and How We Can Escape It*, London, 2015 edn

Harvey A. Levenstein, *Revolution at the Table: the Transformation of the American Diet*, New York, 1988

Robert Lustig, *Fat Chance – The Hidden Truth about Sugar*, New York, 2013

April Merleaux, *Sugar and Civilisation. American Empire and the Cultural Politics of Sweetness*, Chapel Hill, 2015

Sidney Mintz, *Sweetness and Power – The Place of Sugar in Modern History*, London, 1985

Katheryn A. Morrison, *English Shops and Shopping*, New Haven, 2003

Michael Moss, *Salt, Sugar, Fat*, London, 2014

Marion Nestle, *Soda Politics – Taking on Big Soda (and Winning)*, New York, 2015

D. Oddy and D. S. Miller, eds, *The Making of the Modern British Diet*, London, 1975

Avner Offer, *The First World War – An Agrarian Interpretation*, Oxford, 1989

Avner Offer, Rachel Pechey, Stanley Ulijaszek, eds, *Insecurity, Inequality, and Obesity in Affluent Societies*, Proceedings of the British Academy, 174, Oxford, 2012

Matthew Parker, *The Sugar Barons – Family, Corruption, Empire and War*, London, 2012

Barry Popkin, *The World is Fat*, New York, 2009

Mark Pendergast, *For God, Country and Coca-Cola – The Unauthorized History of the Great North American Soft Drink and the Company that Makes it*, New York, 1993

Tsugitaka Sato, *Sugar in the Social Life of Medieval Islam*, Leiden, 2015

Stuart B. Schwarz, *Sugar Plantations in the Formation of Brazilian Society, Bahia, 1550-1835*, Cambridge, 1995

L. D. Schwarz, *London in the Age of Industrialisation*, Cambridge, 1992

Andrew F. Smith, *Drinking History: Fifteen Turning Points in the Making of American Beverages*, New York, 2013

Andrew F. Smith, ed., *The Oxford Encyclopedia of Food and Drink in America*, New York, 2007

Frederick H. Smith, *Caribbean Rum – A Social and Economic History*, Gainesville, Florida, 2005

Jon Stobart, *Sugar and Spice: Grocers and Groceries in Provincial England, 1650– 1830*, Oxford, 2012

Jon Stobart, *Spend, Spend, Spend – A History of Shopping*, Stroud, 2008

Joan Thirsk, *Food in Early Modern England – Phases, Fads, Fashions, 1500–1760*, London, 2007

Jennifer Jensen Wallach, *How America Eats – A Social History of US Food and Culture*, Lanham, Maryland, 2013

James Walvin, *Crossings – Africa, the Americas and the Atlantic Slave Trade,* London, 2013

David Watts, *The West Indies – Patterns of Development, Culture and Environmental Change since 1492*, Cambridge, 1987

Barbara Ketcham Wheaton, *Savoring the Past: the French Kitchen and Table from 1300–1789*, London, 1983

Bee Wilson, *First Bite – How We Learn to Eat,* London, 2015

Wendy A. Woloson, *Refined Tastes – Sugar, Confectionery and Consumers in Nineteenth-Century America*, Baltimore, 2002

John Yudkin, *Pure, White and Deadly*, London, 1972

Numbered References

Introduction – Sugar in Our Time

1 *International Business Review*, 5 September 2016
2 *Sugar Reduction: The Evidence for Action*, Public Health England, London, October 2016

1 – A Traditional Taste

1 Hattie Ellis, 'Honey', in *The Oxford Companion to Sugar and Sweets*, Darra Goldstein, ed., New York, 2015, pp. 336–340
2 Rachel Laudan, *Cuisine and Empire: Cooking in World History*, Los Angeles, 2013, pp. 136–138
3 Nawal Nasrallah, 'Islam', in *The Oxford Companion to Sugar and Sweets*, pp. 361–362
4 Rachel Laudan, *Cuisine and Empire: Cooking in World History*, Los Angeles, 2013, p. 143
5 Sidney Mintz, 'Time, Sugar and Sweetness', in *Food and Culture – A Reader*, Carole Counihan and Penny Van Esterik, eds, London 1997, p. 358
6 Tsugitaka Sato, *Sugar in the Social Life of Medieval Islam*, Boston, 2015, p. 1

7 Tsugitaka Sato, *Sugar in the Social Life of Medieval Islam*, p. 7

8 Tsugitaka Sato, *Sugar in the Social Life of Medieval Islam*, p. 3

9 Tsugitaka Sato, *Sugar in the Social Life of Medieval Islam*, pp. 9–10

10 Peter Brears, *Cooking and Dining in Medieval England*, London, 2008, p. 343

11 Jon Stobart, *Sugar and Spice Grocers and Groceries in Provincial England, 1650–1830*, Oxford, 2012. p. 30

12 Elizabeth Abbott, *Sugar – A Bittersweet History*, London, 2009, p. 20

13 Joan Thirsk, *Food in Early Modern England – Phases, Fads, Fashions, 1500–1760*, London, 2007, pp. 10, 324–325

14 Peter Brears, *Cooking and Dining in Medieval England*, p. 27

15 Peter Brears, *Cooking and Dining in Medieval England*, pp. 344, 379–380, 453–457

16 Peter Brears, *Cooking and Dining in Medieval England*, pp. 453–457

17 Sidney Mintz, *Sweetness and Power – The Place of Sugar in Modern History*, London, 1985, p. 88

18 Ivan Day, 'Sugar Sculptures', *The Oxford Companion to Sugar and Sweets*, pp. 689–693

19 Barbara Ketham Wheaton, *Savoring the Past: the French Kitchen and Table from 1300 to 1789*, London, 1989, pp. 18–21

20 Jon Stobart, *Sugar and Spice*, p. 25

21 Barbara Ketham Wheaton, *Savoring the Past*, pp. 183–184

22 Barbara Ketham Wheaton, *Savoring the Past*, pp. 51–52

23 Barbara Ketham Wheaton, *Savoring the Past*, p. 186

24 Elizabeth Abbott, *Sugar – A Bittersweet History*, p. 22

25 Sidney Mintz, *Sweetness and Power* pp. 90–91

26 Ivan Day, 'Sugar Sculptures', *The Oxford Companion to Sugar and Sweets*, p. 691

27 Ivan Day, 'Sugar Sculptures', *The Oxford Companion to Sugar and Sweets*, pp. 692

28 Jon Stobart, *Sugar and Spice*, pp. 26–27: 30: 56

29 Gervase Markham, *The English Housewife: Containing the Inward and Outward Virtues Which Ought to Be in a Complete Woman*, 1616, edited by Michael R. Best, Montreal, 1986 edition

30 Gervase Markham, *The English Housewife,* Michael R. Best, ed. pp. 72–74, 81–86, 93–94, 103

31 Roy Porter, *The Greatest Benefit to Mankind – A Medical History of Humanity from Antiquity to the Present,* London, 1997, pp. 92–103

32 Roy Porter, *The Greatest Benefit to Mankind,* p. 97

33 Colin Spence, *British Food: An Extraordinary Thousand Years of History,* London, 2002, pp. 48–49; Penelope Hunting, *A History of the Society of Apothecaries,* London, 1988, pp. 18–19

34 Pierre Pomet, *A complete history of drugs. Written in French by Monsieur Pomet. Chief Druggist to the late French King Lewis XIV,* London, 1748, pp. 56–60

2 – The March of Decay

1 Stephen Alforp, 'On a Par with Nixon', *London Review of Books,* 17 November 2016, p. 40; Gervase Markham, *The English Housewife* (1631), Michael R. Best, ed., pp. xxvi, xxxviii, xlii.

2 *The Italian Tribune,* 11 November 2015; *The Daily Telegraph,* 30 September 2016

3 See studies by W. J. Moore and E. Elizabeth Corbett, in *Caries Research,* Klaus G. Konig, ed., vol. 5 1971; vol. 7 1973; vol. 9 1975 and vol. 10 1976. Thanks to Dr Adam Middleton for providing these articles.

4 Neil Walter Kerr, *Dental Caries, Periodontal Disease and Dental Attrition – Their Role in Determining the Life of Human Dentition in Britain over the Last Three Millennia,* Thesis for Doctorate of Medicine, University of Aberdeen, 1999, p. 121

5 *The Smithsonian,* 7 October 2015

6 Colin Jones, *The Smile Revolution in Eighteenth Century Paris,* Oxford, 2014, pp. 9, 17–21, 116

7 Few contemporaries were more troubled by rotting teeth than George Washington. His various dentures offer evidence of his lifetime's struggle to mask his collapsing facial features. His dentures are on display at his home at Mount Vernon, Virginia.

8 B. W. Higman, *A Concise History of the Caribbean,* Cambridge, 2011, p. 104

9 Fernand Braudel, *Capitalism and Material Life, 1400–1800*, London, 1967, pp. 186–188.

10 *The Times*, 20 March 2015

11 'Sharp increase in children admitted to hospital for tooth extract due to decay', *News and Events*, Royal College of Surgeons, 26 February 2016

12 Press Release, 'Tooth decay among 5-year-olds continues significant decline,' Public Health England, 10 May 2016

3 – Sugar and Slavery

1 Stuart B. Schwarz, *Sugar Plantations in the Formation of Brazilian Society, Bahia, 1550–1835*, Cambridge, 1985, pp. 7–9. An arroba = *c.*35lb, making a total of 700,000 lb

2 Stuart Schwarz, *Sugar Plantations*, p. 13

3 James Walvin, *Crossings – Africa, the Americas and the Atlantic Slave Trade*, London, 2013, pp. 35–37

4 Stuart B. Schwarz, *Sugar Plantations*, p. 14

5 David Eltis and David Richardson, *The Atlas of the Transatlantic Slave Trade*, New Haven, 2010, Table 6, p. 202

6 Jonathan Israel, *The Dutch Republic*, Oxford, 1988 edn, p. 116; Sidney Mintz, *Sweetness and Power*, p. 45; Fernand Braudel, *The Wheels of Commerce*, London 1982, p. 193

7 Stuart B. Schwarz, *Sugar Plantations*, pp. 16–22

8 B. W. Higman, *A Concise History*, pp. 97–98

9 B. W. Higman, *A Concise History*, pp. 102–105

10 David Eltis and David Richardson, *Atlas*, Table 6, p. 201

11 For an excellent account, see B. W. Higman, *A Concise History*, Chapter 4, 'The Sugar Revolution'.

4 – Environmental Impact

1 Quoted in David Watts, *The West Indies*, Cambridge, 1987, p. 78

2 B. W. Higman, *A Concise History*, pp. 49–53

3 Quoted in David Watts, *The West Indies*, p. 78

4 David Watts, *The West Indies*, pp. 184–186

5 David Eltis and David Richardson, *Atlas*, Table 6, p. 200

6 David Watts, *The West Indies*, pp. 219–223

7 B. W. Higman, *Jamaica Surveyed*, Kingston, 1988, pp. 8–16

8 David Eltis and David Richardson, *Atlas*, Table 6, p. 200

9 Richard Ligon, *A True and Exact History of the Island of Barbados – 1673*, p. 46.

10 B. W. Higman, *A Concise History*, p. 164

11 B. W. Higman, *A Concise History*, p. 99

5 – Shopping for Sugar

1 Katheryn A. Morrison, *English Shops and Shopping*, New Haven, 2003, p. 5; Jon Stobart, *Spend, Spend, Spend – A History of Shopping*, Stroud, 2008, p. 24

2 Jon Stobart, *Spend*, pp. 25–26

3 Katheryn A. Morrison, *English Shops*, p. 5

4 Jon Stobart, *Spend*, p. 28

5 Jon Stobart, *Sugar and Spice*, pp. 30–31

6 Katheryn A. Morrison, *English Shops*, p. 80

7 Jon Stobart, *Sugar and Spice*, pp. 26–27, 56

8 Jon Stobart, *Sugar and Spice*, p. 114

9 Jon Stobart, *Sugar and Spice*, pp. 72–73

10 James Walvin, *The Quakers – Money and Morals*, London, 1997

11 Jon Stobart, *Sugar and Spice*, p. 118

12 Jon Stobart, *Sugar and Spice*, pp. 170–173

13 *York Courant*, 7 January 1766. (My thanks to Dr Sylvia Hogarth for this reference.)

14 Jon Stobart, *Sugar and Spice*, p. 220

15 Jon Stobart, *Sugar and Spice*, pp. 194–198

16 *Catalogue du Musée de la Sociétié de Pharmacie du Canton de Genève*, n.d. (Wellcome Library, London.)

17 James Walvin, *Slavery in Small Things – Slavery and Modern Cultural Habits*, Chichester, 2017, Chapter 1

6 – A Perfect Match for Tea and Coffee

1 *The Cambridge World History of Food*, Kenneth F. Kiple and Kriemhild Conee Ornelas, eds, 2 vols, Cambridge, 2010, I, p. 647

2 Markman Ellis, Richard Coulton, Matthew Mauger, *Empire of Tea – The Asian Leaf that Conquered the World*, London, 2015, p. 26

3 Markman Ellis, *et al.*, *Empire of Tea*, pp. 43–46

4 Markman Ellis, *et al.*, *Empire of Tea*, p. 56

5 Markman Ellis, *et al.*, *Empire of Tea*, p. 120

6 Kelley Graham, *Gone to the Shops: Going Shopping in Victorian England*, London, 2008, p. 72

7 Quoted in Markman Ellis, *et al.*, *Empire of Tea*, p. 187

8 Rachel Laudan, *Cuisine and Empire – Cooking in World History*, Los Angeles, 2015, p. 229

9 Sidney Mintz, *Sweetness and Power*, pp. 112–114

10 B.W. Higman, *A Concise History*, p. 104

11 Vic Gatrell, *The First Bohemians*, London, 2014 edn, p. 178

12 Markman Ellis, *The Coffee House – A Cultural History*, London, 2004, pp. 79–81

13 Fernand Braudel, *Capitalism and Material Life, 1400–1800*, London, 1967, pp. 186–188

14 Markman Ellis, *The Coffee House*, pp. 202–203

15 Andrew F. Smith, *Drinking History*, New York, 2013, p. 235

16 Jennifer Jensen Wallach, *How America Eats – A Social History of US Food and Culture*, Lanham, Maryland, 2013, pp. 48–49

17 Andrew F. Smith, *Drinking History*, p. 235

18 Richard J. Hooker, *Food and Drink in America – A History*, New York, 1981, p. 130

19 Andrew F. Smith, *Drinking History*, p. 235–236

20 Steven C. Topik, 'Coffee', in Kenneth F. Kiple and Kriemhild Conee Ornelas, eds, *The Cambridge World History of Food*, Cambridge, 2002, vol. I, pp. 644–647

21 Steven C. Topik, 'Coffee', *The Cambridge World History of Food*, pp. 646–647

22 Linda Civitello, *Cuisine and Culture*, Hoboken, 2008, p. 215

23 Richard J. Hooker, *Food and Drink*, p. 201

24 Richard Follett, *The Sugar Masters: Plantations and Slaves in Louisiana's Cane World, 1820–1860*, Baton Rouge, 2005, pp. 20–21

25 Harvey Levenstein, *Revolution at the Table: the Transformation of the American Diet*, New York, 1988, pp. 256–257, n. 2

7 – Pandering to the Palate

1 Rachel Laudan, *Cuisine and Empire*, pp. 185–186
2 David Gentilcore, *Food and Health in Early Modern Europe – Diet, Medicine and Society, 1450-1800*, London, 2015, p. 175
3 Jane Levi, 'Dessert', *The Oxford Companion to Sugar and Sweets*, pp. 211–222
4 Elizabeth Abbott, *Sugar*, pp. 42–49
5 B. W. Higman, *How Food Made History*, pp. 169–172
6 Markman Ellis, *et al.*, *Empire of Tea*, Chapter 9
7 Elizabeth Abbott, *Sugar*, pp. 65–66
8 Sidney Mintz, *Sweetness and Power*, p. 64
9 Carole Shammas, *The Pre-Industrial Consumer in England and America*, Oxford, 1990, pp. 62–66
10 Jessica B. Harris, 'Molasses', in *The Oxford Companion to Sugar*, p. 459

8 – Rum Makes Its Mark

1 Richard Foss, 'Rum', *Oxford Companion to Sugar and Sweets*, pp. 581–582
2 Frederick H. Smith, *Caribbean Rum – A Social and Economic History*, Gainesville, Florida, 2005, pp. 12–15
3 Matthew Parker, *The Sugar Barons – Family, Corruption, Empire and War*, London, 2011, pp. 82–87
4 Kenneth Morgan, *Bristol and the Atlantic Trade in the 18th Century*, Cambridge, 1993, pp. 97–98, 185
5 Roger Norman Buckley, *The British Army in the West Indies – Society and Military in the Revolutionary War*, Gainesville, 1998, pp. 284–285
6 Frederick H. Smith, *Caribbean Rum*, p. 21
7 Richard Hough, *Captain James Cook – A Biography*, London, 1994, p. 67
8 Frederick H. Smith, *Caribbean Rum*, p. 28
9 Jacob M. Price, 'The Imperial Economy', in *The Oxford History of the British Empire*, vol. II, P. J. Marshall, ed., p. 90
10 Thomas Bartlett, 'Ireland and the British Empire', in *The Oxford History of the British Empire*, vol. II, P. J. Marshall, ed., p. 257

11 Wendy A. Woloson, *Refined Tastes – Sugar, Confectionery and Consumers in Nineteenth Century America*, Baltimore, 2002, pp. 17, 23–26

12 Wendy Woloson, *Refined Tastes*, p. 5

13 Frederick H. Smith, *Caribbean Rum*, pp. 29–30

14 Thanks to James Axtell for this point.

15 Frederick H. Smith, *Caribbean Rum*, pp. 76–81; Andrew F. Smith, *Drinking History*, pp. 26–28

16 Frederick H. Smith, *Caribbean Rum*, pp. 81–86; B. W. Higman, *Slave Populations and Economy in Jamaica*, p. 21

17 A. G. L. Shaw, *Convicts and Colonies*, London, 1966, pp. 66–68; Charles Bateson, *The Convict Ships, 1787-1868*, Glasgow, 1959, pp. 66–67

18 Deidre Coleman, ed., *Maiden Voyages and Infant Colonies*, London, 1999, p. 126

19 David Eltis, *The Rise of African Slavery in the Americas*, Cambridge, 2000, pp. 127–128; Frederick H. Smith, *Caribbean Rum*, p. 99

20 Michael Craton and James Walvin, *A Jamaican Plantation – Worthy Park, 1670–1970*, London, 1970, p. 136

9 – Sugar Goes Global

1 Paul Dickson, 'Combat Food', in Andrew F. Smith, *The Oxford Companion to American Food and Drink*, New York, 2007, pp. 141–142

2 British Army Rations, Vestey Foods. At https://worldwarsupplies. co.uk; accessed 5 April 2016

3 *US Populations 1776 to Present* – https://fusiontables.google.com

4 P. Lynn Kennedy and Won W. Koo, eds, *Agricultural Trade Policies in the New Millennium*, New York, 2006, p. 156

5 B. W. Higman, *How Food Made History*, Chichester, 2012, pp. 67–68

6 James Walvin, *Atlas of Slavery*, London, 2006, p. 123

7 P. Lynn Kennedy and Won W. Koo, eds., *Agricultural Trade Policies*, p. 157

10 – The Sweetening of America

1 Harvey Levenstein, *Revolution at the Table*, pp. 32–33
2 Richard Sutch and Susan B. Carter, eds., *Historical Statistics of the United States*, Cambridge, 2006, pp. 553–555
3 Harvey Levenstein, *Revolution at the Table*, pp. 32–33
4 Jennifer Jensen Wallach, *How America Eats*, p. 71
5 Linda Civitello, *Cuisine and Culture*, p. 210
6 Reginal Horsman, *Feast or Famine – Food and Drink in American Westward Expansion*, Columbia, Missouri, 2000, p. 302
7 'Wedding Cakes', *The Oxford Companion to American Food and Drink*, Andrew F. Smith, ed., p. 618
8 'Fruit Preserves', in *Oxford Companion to Sugar and Sweets*, p. 282
9 Jennifer Jensen Wallach, *How America Eats*, p. 100
10 Andrew F. Smith, *Pure Ketchup – A History of America's National Condiment*, Columbia, South Carolina, Ch. 3
11 Jennifer Jensen Wallach, *How America Eats*, pp. 101–104

11 – Power Shifts in the New World

1 Michael Duffy, *Soldiers, Sugar and Seapower*, Oxford, 1987, pp. 7–13
2 Elizabeth Abbott, *Sugar*, pp. 180–181
3 B. W. Higman, *A Concise History*, p. 166
4 Elizabeth Abbott, *Sugar*, pp. 181–183
5 Ronald Findlay and Kevin H. O'Rourke, *Power and Plenty*, Princeton, 2007, pp. 366–369
6 Franklyn Stewart Harris, *The Sugar Beet in America*, New York, 1919, Chapter II
7 Quoted in Gail M. Hollander, *Raising Cane in the 'Glades: the Global Sugar Trade and the Transformation of Florida*, Chicago, 2008, p. 46
8 Richard Follett, *The Sugar Masters*, pp. 30–31
9 Richard Follett, *The Sugar Masters*, p. 24
10 Richard Follett, *The Sugar Masters*, p. 27
11 William Ivey Hair, *Bourbonism and Agricultural Protest – Louisiana*

Politics, 1877–1900, Baton Rouge, 1969, pp. 38–39

12 A. B. Gilmore, 'Louisiana Sugar Manual', New Orleans, 1920, Table 1; Fargo, N.D.

13 David Eltis and David Richardson, *Atlas*, p. 202

14 James Walvin, *Crossings*, pp. 185–187

15 James Walvin, *Crossings*, p. 204

16 Elizabeth Abbott, *Sugar*, pp. 273–280

17 Alfred Eichner, *The Emergence of Oligarchy – Sugar Refining as a Case Study*, Baltimore, 1966, pp. 339–342

18 Cesar J. Ayala, *American Sugar Kingdom: The Plantation Economy of the Spanish Caribbean*, Chapel Hill, 1999, p. 3; April Merleaux, *Sugar and Civilisation – American Empire and the Cultural Politics of Sweetness*, Chapel Hill, 2015

19 Jacob Adler, *Claus Speckels, The Sugar King in Hawaii*, Honolulu, 1966

20 Jacob Adler, *Claus Spreckels, The Sugar King*, p. 205

21 Elizabeth Abbott, *Sugar*, Chapter 10

22 Cesar J. Ayala, *American Sugar Kingdom*

23 Cesar J. Ayala, *American Sugar Kingdom*, p. 5

24 Maxey Robson Dickson, *The Food Front in World War I*, Washington DC, 1944, pp. 11–12, 25

25 Maxey Robson Dickson, *The Food Front*, pp. 148–149

26 Maxey Robson Dickson, *The Food Front*, p. 149

27 Maxey Robson Dickson, *The Food Front*, pp. 150–151

28 Cindy Hahamovitch, *No Man's Land : Jamaican Guestworkers in America and the Global History of Deportable Labor*, Princeton, 2011, p. 138

29 Cindy Hahamovitch, *No Man's Land*, p. 138

30 Cindy Hahamovitch, *No Man's Land*, p. 139

31 Cindy Hahamovitch, *No Man's Land*, p. 141

32 Cindy Hahamovitch, *No Man's Land*, p. 142

33 Cindy Hahamovitch, *No Man's Land*, pp. 3–7

34 Alex Wilkins, *Big Sugar – Seasons in the Cane Fields of Florida*, New York, 1989, p. 49

35 Dexter Filkins, 'Swamped', in *The New Yorker*, 4 January 2016

36 Gail M. Hollander, *Raising Cane*, pp. 42–46

12 – A Sweeter War and Peace

1 Avner Offer, *The First World War: An Agrarian Interpretation*, Oxford, 1984, p. 297

2 L. D. Schwarz, *London in the Age of Industrialisation*, Cambridge, 1992, p. 41

3 G. N. Johnson, 'The Growth of the Sugar Trade and Refining Industry', in D. Oddy and D. S. Miller, eds, *The Making of the Modern British Diet*, London, 1975, Ch. 5

4 Henry Weatherley, *Treatise on the Art of Boiling Sugar*, 1864

5 George Dodd, *The Food of London*, London, 1856, p. 428

6 Ben Fine, Michael Heasman and Judith Wright, *Consumption in the Age of Affluence: The World of Food*, London 1996, p. 96

7 Avner Offer, *The First World War*, p. 39

8 Avner Offer, *The First World War*, pp. 39, 168

9 G. N. Johnson, 'The Growth of the Sugar Trade and Refining Industry', D. Oddy and D. S. Miller, eds., *The Making of the Modern British Diet*, pp. 60–61

10 Ben Fine, *et al.*, *Consumption*, pp. 94–95

11 Ben Fine, *et al.*, *Consumption*, p. 99

12 Peter Mathias, *Retailing Revolution*, London, 1967, p. 56

13 Stuart Thorpe, *The History of Food Preservation*, Kirby Lonsdale, 1986, p. 152; Sue Shepherd, *Pickled, Potted and Canned – The Story of Food Preserving*, London, 2000, p. 164

14 Peter Mathias, *Retailing Revolution*, pp. 103–104

15 Helen Franklin, 'As Good as Five Shillings a Week – Poor Dental Health and the Establishment of Dental Provision for Schoolchildren in Edwardian England', MA thesis, University of London [Wellcome Library], pp. 15–17, Thomas Oliver, 'Our Workmen's Diet and Wages', *The Fortnightly Review*, vol. 56, October 1899, p. 519

16 Ben Fine, *et al.*, *Consumption*, pp. 100–101

17 James Walvin, *The Quakers*, Ch. 10

18 Helen Franklin, 'As Good as Five Shillings a Week . . .', pp. 15–17; Ben Fine et al., *Consumption* . . ., pp. 99–101

19 Ben Fine *et al.*, *Consumption* . . ., pp. 101–102

20 G. N. Johnson, 'The Growth of the Sugar Trade and Refining Industry', D. Oddy and D. S. Miller, eds, *The Making of the Modern British Diet*, p. 60

21 L. Margaret Barnett, *British Food Policy During the First World War*, London, 1985, pp. 30–31

22 L. Margaret Barnett, *British Food Policy . . .*, p. 138

23 Ben Fine, *et al.*, *Consumption . . .*, p. 96

24 Robert Graves, *Goodbye to All That*, London, 2000 edn, p. 82

25 Ben Fine, *et al.*, *Consumption . . .*, p. 102

26 Ben Fine, *et al.*, *Consumption . . .*, pp. 101–103

27 Ben Fine, *et al.*, *Consumption . . .*, pp. 103–104

28 B. Kathleen Hey, *The View from the Corner Shop – The Diary of a Yorkshire Shop Assistant in Wartime*, Patricia and Robert Malcolmson, eds, London, 2016, p. 119

29 Ben Fine, *et al.*, *Consumption . . .*, pp. 122–123

30 Ron Noon, 'Goodbye, Mr Cube', *History Today*, Vol, 51, No. 10, October 2007

31 Ben Fine, *et al.*, *Consumption . . .*, pp. 103–104

13 – Obesity Matters

1 Iona and Peter Opie, *The Lore and Language of Schoolchildren*, Oxford, 1959, pp. 167–169; Francis Delpeuch and others, *Globesity: a Planet out of Control?*, London, 2009, p. 30

2 For a general discussion, see David Haslam and Fiona Haslam, *Fat, Gluttony and Sloth: Obesity in Medicine, Art and Literature*, Liverpool, 2009

3 Francis Delpeuch, *Globesity*, pp. 43–44

4 David Lewis and Margeret Leitch, *Fat Planet*, p. xv

5 David Lewis and Margaret Leitch, *Fat Planet*, p. xi. For a recent scholarly study of the problem, see *Insecurity, Inequality, and Obesity in Affluent Societies*, edited by Avner Offer, Rachel Pechey, Stanley Ulijaszek, *Proceedings of the British Academy*, 174, Oxford, 2012

6 'NHS spending millions on larger equipment for obese patients', *Guardian*, 24 October 2015

7 Cathy Newman, 'Why are we so fat?' *National Geographic*, nationalgeographic.com/science/health. Accessed 27 July 2016

8 David Lewis and Margaret Leitch, *Fat Planet*, p. xii

9 Four Decade Study: 'Americans Taller, Fatter, by Live Science Staff', 27 October 2004, www/livescience.com; accessed 28 July 2016

10 George Vigarello, *The Metamorphoses of Fat – a History of Obesity*, Columbia University Press, 2013, p. 186

11 Nana Bro Folman *et al.*, 'Obesity, hospital service use and costs', in Kristian Bolin and John Cawley, eds, *The Economics of Obesity*, Amsterdam, 2007, p. 329; 'Americans are still Getting Fatter', Vice News: www//news.vice.co/article; accessed 28 July 2016; Tyler Durden, 'Americans Have Never Been Fatter', 'Healthcare costs attributable to Obesity', Zero hedge, www zerohedge.com; accessed 28 July 2016

12 Julie Lumeng, 'Development of Eating Behaviour,' in *Obesity: Causes, Mechanisms, Prevention, and Treatment*, Elliott M. Blass, ed., Sunderland, MA, 2008

13 'How Americans Got Fat', in charts, www.bloomberg.com/news/articles 2016; accessed 28 July 2016

14 Michael Moss, *Salt, Sugar, Fat: How the Food Giants Hooked Us*, London, 2014, p. 22

15 Francis Delpeuch, *Globesity*, p. 10; David Lewis and Margaret Leitch, *Fat Planet*, pp. xii–xiii

16 David Lewis and Margaret Leitch, *Fat Planet*, pp. xii–xiii

17 George Vigarello, *The Metamorphosis of Fat . . .*, p. 186

18 'NHS Choices: Your Health, Your Choices', www.nhs.uk/Liveweight; accessed 28 June 2016

19 Foresight; 'Tackling Obesity: Future Choices', *Project Report*, London, 2007, p. 5

20 The *Guardian*, 23 October 2015

21 Francis Delpeuch, *Globesity*, Chapter 1

22 David Lewis and Margaret Leitch, *Fat Planet*, p. xiv

23 David Crawford and Robert W. Jeffery, *Obesity Prevention and Public Health*, Oxford, 2015, pp. 17, 212

24 David Crawford and Robert W. Jeffery, *Obesity Prevention and Public Health*, pp. 17, 212

25 Michael Gard and Jan Wright, *The Obesity Epidemic: Science, Morality and Ideology*, London, 2005, pp. 3–6. This book comes close to arguing that obesity is a moral panic.

26 More recently, however, there seems to have been a levelling-off of Western childhood obesity – though the problem persists among low-income groups. Francis Delpeuch, *Globesity*, pp. 13–14

27 Francis Delpeuch, *Globesity*, pp. 15–16

28 Francis Delpeuch, *Globesity*, pp. 15–16

29 Jeffrey P. Koplin, Catheryn T. Liverman, Vivica I Kraak, eds, *Preventing Childhood Obesity: Health in the Balance*, Washington, DC, 2005, pp. xiii, 21–22, 73

30 Michael Moss, *Salt, Sugar, Fat*, pp. 22–23

31 David Crawford and Robert W. Jeffery, *Obesity Prevention and Public Health*, pp. 15–16

32 Naveed Sattar and Mike Lean, eds, *ABC of Obesity*, BMJ Books, Blackwell, Oxford, 2007, p. 38

33 'The Junk Food Toll', the *Guardian*, 8 October 2016

34 Julie Lumeng, 'Development of Eating Behaviour . . .' in Elliott M. Blass, ed, *Obesity*, p. 163

35 'The State of Children's Oral Health in England', January 2015, Faculty of Dental Surgery, Royal College of Surgeons

36 'The State of Children's Oral Health in England', January 2015, Faculty of Dental Surgery, Royal College of Surgeons, pp. 3–5

37 'The State of Children's Oral Health in England', January 2015, Faculty of Dental Surgery, Royal College of Surgeons, pp. 5–6

38 'The State of Children's Oral Health in England', January 2015, Faculty of Dental Surgery, Royal College of Surgeons, p. 4

39 'The State of Children's Oral Health in England', January 2015, Faculty of Dental Surgery, Royal College of Surgeons, p. 7

40 Haroon Siddique, 'Children eating equivalent of 5,500 sugar lumps a year', the *Guardian*, 4 January 2016

41 Frances M. Berg, *Underage and Overweight*, New York, 2005, p. 102

42 'Sugars and tooth decay', www.actiononsugar.org; accessed 28 August 2016

43 'Which foods and drinks containing sugar cause tooth decay?'

NHS Choices, *Your Health, Your Choice*, www.nhs.uk/which-foods-and-drinks; accessed 28 August 2016

44 NHS Choices, www.nhs/news/2015/03March; accessed 13 January 2015

45 Hilary Lawrence, *Not on the Label: What Really Goes into Food on Your Plate*, London, 2004, pp. 220–223

46 Shauna Harrison, Darcy A. Thompson, Dina L. G. Borzekowski, 'Environmental Food Messages . . .', Chapter 12, in Elliott M. Blass, ed., *Obesity* pp. 372–392

47 Bee Wilson, *First Bite – How We Learn to Eat*, London, 2015, pp. 86–89

48 Francis Delpeuch, *Globesity*, pp. 23–26

49 Francis Delpeuch, *Globesity*, pp. 1–3

14 – The Way We Eat Now

1 Barry Popkin, *The World is Fat*, New York, 2009, p. 30

2 Michael Moss, *Salt, Sugar, Fat*, p. 11

3 Michael Moss, *Salt, Sugar, Fat*, p. 15

4 Michael Moss, *Salt, Sugar, Fat*, pp. 63–64

5 Michael Moss, *Salt, Sugar, Fat*, pp. 70–71

6 Michael Moss, *Salt, Sugar, Fat*, pp. 55–56

7 Michael Moss, *Salt, Sugar, Fat*, pp.72–74

8 Michael Moss, *Salt, Sugar, Fat*, pp. 80–82

9 Felicity Lawrence, *Not on the Label*, p. 268

10 *The Wall Street Journal*, 29 July 2015

11 Felicity Lawrence, *Not on the Label*, pp. 268–269

12 Quoted in Felicity Lawrence, *Not on the Label*, p. 273

13 Derek Oddy and Alain Drouard, eds., *The Food Industries of Europe in the 19th and 20th Centuries*, Farnham, 2013, p. 5

14 Derek Oddy and Alain Drouard, eds, *The Food Industries of Europe* pp. 240–245

15 This issue is well explained in Michael Moss, *Salt, Sugar, Fat*.

16 Michael Moss, *Salt, Sugar, Fat*, p. xxvi

17 David Lewis and Margaret Leitch, *Fat Planet*, pp. 210, 53

18 Quoted in David Lewis and Margaret Leitch, *Fat Planet*, p. 210

19 Michael Moss, *Salt, Sugar, Fat*, pp. 10–11

20 Michael Moss, *Salt, Sugar, Fat*, p. 11

21 David Lewis and Margaret Leitch, *Fat Planet*, Chapter 7

22 'How the Sugar Industry Shifted Blame to Fat', *New York Times*, 12 September 2016

23 The most prominent victim was John Yudkin and his pioneering book, *Pure, White and Deadly*, London, 1972

24 Alice P. Julier, 'Meals', in Anne Murcott, Warren Bell, Peter Jackson, eds, *The Handbook of Food Research*, London, 2013

25 Francis Delpeuch, *Globesity*, 2009, p. 37

26 Francis Delpeuch, *Globesity*, pp. 37–38

27 Michael Moss, *Salt, Sugar, Fat*, p. 246

28 Isabelle Lescent Giles 'The Rise of Supermarkets in 20th Century Britain and France', in *Land, Shops and Kitchen*, Carmen Sarasua, Peter Schollier and Leen Van Molle, eds, Turnhout, Belgium, 2005, Ch. 10

29 David Lewis and Margaret Leitch, *Fat Planet*, p. 215

30 Louise O. Fresco, *Hamburgers in Paradise: the Stories Behind the Food We Eat*, Princeton, 2016, pp. 325-328

31 David Lewis and Margaret Leitch, *Fat Planet*, p. 202

32 Neil Pennington and Charles W. Baker, eds, *Sugar – A User's Guide to Sucrose*, New York, 1990, p. 103

33 Neil Pennington and Charles W. Baker, eds, *Sugar*, pp. 165, 171, 177

15 – Hard Truth About Soft Drinks

1 Colin Emmins, 'Soft Drinks', in Kenneth F. Kiple and Kriemhild Conee Ornelas, eds, *The Cambridge World History of Food*, Cambridge, 2000, vol. I, pp. 702–711

2 Andrew Coe, 'Soft Drinks', in *The Oxford Companion to American Food and Drink*, Andrew F. Smith, ed., New York, 2007, p. 546

3 Andrew Coe, 'Soft Drinks', in *The Oxford Companion to American Food and Drink*, Andrew F. Smith, p. 54

4 Andrew Coe 'Soft Drinks' in *The Oxford Companion to American Food and Drink*, Andrew F. Smith, ed., p. 546

5 Joseph E. McCann, *Sweet Success: How NutraSweet Created a Billion Dollar Business*, Illinois, 1990, p. 80

6 Bartow J. Elmore, *Citizen Coke – The Making of the Coca-Cola Capitalism*, New York, 2010, pp. 76–77

7 Bartow J. Elmore, *Citizen Coke* . . ., pp. 85–87

8 Bartow J. Elmore, *Citizen Coke* . . ., pp. 100–101

9 Bartow J. Elmore, *Citizen Coke* . . ., pp. 105–106

10 Bartow J. Elmore, *Citizen Coke* . . ., pp. 106–107

11 Mark Prendergast, *For God, Country and Coca-Cola: The Unauthorized History of the Great North American Soft Drink and the Company that Makes It*, New York, 1993, pp. 199, 203–204

12 Bartow J. Elmore, *Citizen Coke* . . ., pp. 158–159, 207

13 Letters quoted are reprinted in Mark Prendergast, *For God and Country*, pp. 210–213

14 Bartow J. Elmore, *Citizen Coke* . . ., pp. 168–169

15 Bartow J. Elmore, *Citizen Coke* . . ., pp. 169–179

16 Bartow J. Elmore, *Citizen Coke* . . ., pp. 186–192

17 Bartow J. Elmore, *Citizen Coke* . . ., pp. 178–179

18 Bartow J. Elmore, *Citizen Coke* . . ., pp. 181–182

19 'How the business of bottled water went mad', the *Guardian*, 6 October 2016

20 Bartow J. Elmore, *Citizen Coke* . . ., pp. 263–264

21 Erik Millstone and Tim Lang, *The Atlas of Food*, Brighton, 2008 edn., p. 91; Joseph E. McCann, *Sweet Success*

22 'Corn Syrup', *The Oxford Companion to Sugar and Sweets*, p. 189; Bartow J. Elmore, *Citizen Coke* . . ., pp. 267–269

23 Andrew I. Smith, *Encyclopedia of Junk Food and Fast Food*, Westport, 2008, pp. 259–260

24 Bartow J. Elmore, *Citizen Coke* . . ., p. 270

25 For a precise analysis of these trends, see John Komlos and Marek Brabec, 'The Transition to Post-Industrial BMI Values in the United States', Chapter 8, in *Insecurity, Inequality, and Obesity in Affluent Societies*, by Avner Offer, Rachel Pechey, Stanley Ulijaszek, eds, *Proceedings of the British Academy*, No. 174, 2012; Bartow J. Elmore, *Citizen Coke* . . ., pp. 270–271

26 Bartow J. Elmore, *Citizen Coke* . . ., pp. 272–273

27 Michael Moss, *Salt, Sugar, Fat*, pp. 97–109

28 Michael Moss, *Salt, Sugar, Fat*, pp. 110–113

29 Michael Moss, *Salt, Sugar, Fat*, pp. 113–115
30 Michael Moss, *Salt, Sugar, Fat*, pp. 116–117
31 Michael Moss, *Salt, Sugar, Fat*, pp. 131–132

16 – Turning the Tide – Beyond the Sugar Tax

1 'The WHO calls on countries to reduce sugar intake among adults and children', WHO Media Centre, 4 March 2015
2 Statistics and facts on Health and Fitness Clubs, www.statistics.com; accessed 28 January 2016; 'State of the UK Fitness Industry', report, June 2015, www.leisured.com; accessed 28 January 2015
3 *Sugar Reduction: The Evidence for Action*, 2015, London, p. 5
4 *Sugar Reduction*, p. 9
5 Avner Offer, *The Challenge of Affluence: Self-Control and Wellbeing in the United States and Britain Since 1950*, Oxford, 2006
6 *Sugar Reduction*, p. 22
7 *Sugar Reduction*, p. 20
8 *Sugar Reduction*, p. 22
9 *Sugar Reduction*, p. 21
10 *Sugar Reduction*, p. 28
11 'Revealed: high sugar content of hot drinks', the *Guardian*, 17 February 2016
12 *Sugar Reduction*, p. 27
13 *Sugar Reduction*, p. 30
14 *Sugar Reduction*, p. 23
15 First Leader, *The Times*, 22 October 2015. Thereafter, major journalists continued in much the same vein.
16 David Aaronovitch, 'We need heavy weapons to win the obesity war', *The Times*, 28 July 2016
17 The *Daily Telegraph*, 3 January 2015
18 Robert H. Lustig, *The Hidden Truth about Sugar, Obesity and Disease,* London, 2014
19 Sarah Knapton, 'Sugar is as dangerous as alcohol and tobacco . . .' The *Daily Telegraph*, 9 January 2014
20 'NHS Chief to introduce sugar tax in hospitals . . .' The *Guardian*, 16 January 2016

21 Tina Rosenberg, 'Mexico's Fat Tax', the *Guardian*, 3 November 2015

22 *The Times*, 7 January 2016

23 Coca-Cola had abandoned cane sugar for fructose corn syrup in 1980

Conclusion – Bitter-Sweet Prospects

1 J. H. Galloway, 'Sugar' in Kenneth F. Kiple and Kriemhild Conee Ornelas, eds, *The Cambridge World History of Food*, 2 vols, vol. II, p. 446; John McQuaid, *Tasty: The Art and Science of What We Eat*, New York, 2015, p. 119

Index